Cognitive Systems Monographs

Volume 22

About this Series

The Cognitive Systems Monographs (COSMOS) publish new developments and advances in the fields of cognitive systems research, rapidly and informally but with a high quality. The intent is to bridge cognitive brain science and biology with engineering disciplines. It covers all the technical contents, applications, and multidisciplinary aspects of cognitive systems, such as Bionics, System Analysis, System Modelling, System Design, Human Motion, Understanding, Human Activity Understanding, Man-Machine Interaction, Smart and Cognitive Environments, Human and Computer Vision, Neuroinformatics, Humanoids, Biologically motivated systems and artefacts Autonomous Systems, Linguistics, Sports Engineering, Computational Intelligence, Biosignal Processing, or Cognitive Materials as well as the methodologies behind them. Within the scope of the series are monographs, lecture notes, selected contributions from specialized conferences and workshops, as well as selected PhD theses.

Jeremy L. Wyatt · Dean D. Petters
David C. Hogg
Editors

From Animals to Robots and Back: Reflections on Hard Problems in the Study of Cognition

A Collection in Honour of Aaron Sloman

 Springer

Editors
Jeremy L. Wyatt
School of Computer Science
University of Birmingham
Birmingham
UK

David C. Hogg
School of Computing
University of Leeds
Leeds
UK

Dean D. Petters
School of Social Sciences
University of Northampton
Northampton
UK

ISSN 1867-4925
ISBN 978-3-319-06613-4
DOI 10.1007/978-3-319-06614-1
Springer Cham Heidelberg New York Dordrecht London

ISSN 1867-4933 (electronic)
ISBN 978-3-319-06614-1 (eBook)

Library of Congress Control Number: 2014939638

Printed on acid-free paper

Springer is part of Springer Science+Business Media (www.springer.com)

Foreword

This collection of papers is based on talks and papers given at a symposium organised to celebrate Aaron Sloman's 75th birthday. The event took place at the University of Birmingham on 12 and 13 September 2011. Approximately 70 attendees came from major AI research centres where Aaron has worked such as Birmingham, Edinburgh and Sussex, as well as from many universities around the world where his former students and collaborators now work. These included colleagues from as far afield as Canada and California. The symposium included both academic talks and a lively dinner where many delegates were able to testify to Aaron's impact on their work to date. A common theme among the speakers was that their academic lives have never been quite the same again since working with him.

Aaron has made a remarkably wide ranging impact in the disciplines of artificial intelligence and cognitive science. He has worked in the fields of artificial intelligence and cognitive science since 1972, when he worked as a Senior Visiting Fellow in Edinburgh for a year. He comments on that time: *I think I learnt more in that year than in any other year of my life since about the age of 4* and that he was *converted to A. I. as the best way to do philosophy* (Sloman 2012, p. 2). His unusually broad interests and papers are detailed on his homepage (Sloman 2014a) and the websites for the Cognition and Affect Project (Sloman 2014b) and the Meta–Morphogenesis Project (Sloman 2014c). Additionally, in this collection, Maggie Boden's contribution provides an overview of his work. He has supervised or worked with many people in the field. From the editors and contributors in this collection are included six former Ph.D. students of Aaron: David Hogg, Tom Khabaza, Tim Read, Luc Beaudoin, Ian Wright, Nick Hawes, and Dean Petters. Also represented are former postdocs: Brian Logan, Matthias Scheutz, and Michael Zillich. Other contributions come from Aaron's current or former faculty colleagues: Maggie Boden, Manfred Kerber, Jeremy Wyatt, and Jeremy Baxter. In addition to his prodigious and wide ranging research output, Aaron has also had a profound influence in shaping two university departments: at Sussex University developing a Cognitive Studies Programme in the School of Social Sciences into the School of Cognitive and Computing Sciences (COGS) and at the University of

Birmingham laying the foundations for successful research and teaching in artificial intelligence and cognitive science. Both of these have interdisciplinary and notably friendly cultures, and these are hallmarks of Aaron's influence.

Jeremy L. Wyatt
Dean D. Petters
David C. Hogg

References

Sloman A (2012) Aaron Sloman—Curriculum Vitae, available online at http://www.cs.bham.ac.uk/axs/cv.pdf

Sloman A (2014a) Aaron Sloman's Homepage. http://www.cs.bham.ac.uk/axs/, accessed: 2014-02-26

Sloman A (2014b) The Cognition and Affect Project. http://www.cs.bham.ac.uk/research/projects/cogaff/0-INDEX.html#contents, accessed: 2014-02-26

Sloman A (2014c) The Meta–Morphogenesis (MM) Project. http://www.cs.bham.ac.uk/research/projects/cogaff/misc/meta-morphogenesis.html, accessed: 2014-02-26

Acknowledgments

We extend our sincere thanks to all the contributors to this collection. They enthusiastically engaged with the spirit of the original symposium, authoring papers on a bewildering range of topics that does at least some justice to the enormous range of Aaron's interests and achievements. The attendees at the symposium itself were of an equally broad range: including not only academics, but also industrialists and entrepreneurs.

We also thank all who helped with the organisation of the symposium. In particular, we would like to extend our thanks to David Lodge and Stephanie Dale who joined us to read a delightful excerpt from a play based on David's novel "Thinks", featuring Professor of Cognitive Science Ralph Messenger engaging in a lively debate with an academic colleague on the nature of cognition. David edited the excerpt especially for the event, and also spoke about how Aaron had helped him with much of the science and philosophy expressed in the book.

We also thank all our families, in particular Dean would like to thank Cath, Lauren and Beth for support during the process of editing this book.

Finally, there is one person to whom we owe the biggest thank you of all. Aaron, you have challenged us, changed us, provoked new ideas, new research programmes and taught us much about how to be better academics. We are much richer for the experience, and this collection is our thank you to you.

Jeremy L. Wyatt
Dean D. Petters
David C. Hogg

Contents

Chapter 1
Bringing Together Different Pieces to Better Understand Whole Minds

Dean Petters

This collection of chapters looks at a diverse set of topics in the study of cognition. These topics do not of course, despite their broad range, include all the pieces needed for a whole mind. However, what this collection provides is explorations of how some very different research topics can be brought together through shared themes and outlook. Each chapter in this collection takes a reflective view of work in one area. Chapters may cover empirical results as well as theories about cognition—in people, other animals and robots. The scope includes what problems there are, or could be solved, what requirements there are to solve those problems, how they might be solved in practice and some predictions about how each field will evolve. The wide spread of the chapters follows the broad interests and research of Aaron Sloman and demonstrates the broad range of subfields in artificial intelligence and cognitive science that need to be considered in the design of intelligent artificial systems and the understanding of natural ones. Much interesting and valuable scholarship in A.I. and cognitive science is focussed on narrow issues—going deep into problems in areas like perception, exploration, planning, automated proof solving, learning, or action selection and motor control. This collection includes such specialist areas of scholarship but brings them together as part of a cross-disciplinary effort to understand natural minds and design artificial minds.

Included amongst the contributors are a range of speakers who have worked with Aaron Sloman in the past, or are working with him now. Maggie Boden and Aaron Sloman were both founder member of the COGS School of Cognitive and Computing Sciences at the University of Sussex. In her contribution to this collection, Boden starts by reviewing the broad sweep of progress in AI from the publication of Turing (1950) Mind paper onwards. The work of Aaron Sloman is then situated within this long view as particularly oriented around philosophical and scientific aims of understanding natural cognition through designing artificial systems. As Boden notes, Sloman throughout his career has attempted to understand 'the mind as a whole'. She

D. Petters (✉)
School of Social Sciences, University of Northampton, Northampton, UK
e-mail: dean.petters@northampton.ac.uk

J. L. Wyatt et al. (eds.), *From Animals to Robots and Back: Reflections on Hard Problems in the Study of Cognition*, Cognitive Systems Monographs 22, DOI: 10.1007/978-3-319-06614-1_1, © Springer International Publishing Switzerland 2014

emphasises Sloman's conceptual heritage from diverse philosophers, most importantly Kant. Boden then goes onto show the continuity between Sloman's explicitly philosophical research and the philosophical motivations for his research, which at first impressions might be removed from these concerns and be more focussed on the engineering aims of A.I., such as in Sloman's vision research with the POPEYE system. From vision research Boden's contribution turns to consider emotion. Whereas perceptual capabilities can be considered at a subpersonal level—many aspects of a whole mind, such as emotion, are better considered as global states of a whole agent or organism. Sloman has demonstrated that understanding emotions such as love (Sloman 2000) and grief (Wright et al. 1996) requires showing how information processing architectures for complete agents can give rise to these emotional states. Boden then reviews Sloman's design-based approach to research on information processing architectures more generally before bringing Sloman's research uptodate with a brief review of his recent work. In the latter part of her chapter, Boden reflects on why recognition for Sloman's work was for a time delayed but is now increasing.

Boden's contribution weaves together many of the disparate interests of Aaron Sloman and highlights the strong underlying links between these interests. The following chapters are more narrow and specialist than this—often confining themselves to just some limited part of Aaron Sloman's range of interests. However, there are several themes which cut across many of the remaining chapters in this collection. All the chapters in the collection take a philosophical view of their particular topic being reflected upon. In particular, the aims of philosophy and science overlap because both disciplines are concerned with considering possibilities, forming new representations of these possibilities, and attempting to explain why the limits of the possible are as they are (Sloman 1978, Chaps. 2 and 3). All the chapters in this collection are thus scientific and philosophical in a similar design-based sense. This means that in trying to explain possibilities in natural and artifical cognition, they consider these issues from the perspective of a designer—how might a designer reproduce cognitive phenomena of interest.

Other themes cut across many of the chapters but are more strongly focused upon in some chapters than others. Petter takes an architectural and design-based approach to explain emotional phenomena. He starts by contrasting shallow and deep models of emotion. Shallow models might explain a relatively narrow set of phenomena of interest—which might be behavioural or phenomenological or involve observations from a biological, cognitive or functional perspective. Whereas deep models arise from attempting to explain a much broader set of phenomena that are chosen to provide a set of requirements that does not underspecify emotion states. Petters then goes onto review explanatory frameworks for emotion research that focus on explaining emotions in terms of particular phases of emotion experience or particular emotional components. This review concludes that an integrative approach such as Sloman's Cogaff Schema is required to explain examples of loss of control in humans which are more substantial, complex and clinically significant than many emotion researchers or clinicians might imagine a modelling approach can cope with. Using the Cogaff approach is beneficial because it naturally integrates all types of information processing across the range of processing phases. The chapter ends by presenting putative explanations of six emotional phenomena that involve loss of control. Since it does so in terms of information processing architectures, it links with many other chapters in this collection.

Chappell reviews the state of the art in research on how animals gather and represent information about the world. In particular, Chappell describes current challenges and future prospects for the study of physical cognition across a broad range of animals and a broad range of tasks. This review includes the contrasting building activities of web building spiders and nest building bower birds, and contrasting perceptual learning skills in Honey bees and parrots. How Orangutans navigate big gaps between trees by using their weight to lean appropriate branches the right way is analysed. Chappell also considers the cognitive abilities that support the tool use of crows and rooks (such as understanding causal roles of materials). These factors are then assessed alongside morphological adaptations that facilitate tool use (like the 'right kind' of straight bill and binocular visual field). Looking to the future, Chappell then highlights areas of expected growth in the understanding of animal cognition including in perception, attention and exploration. From consideration of animal cognition the collection then jumps to three chapters by respectively: Kerber; Logan; and Scheutz; that consider information processing within architectures within a formal mathematical approach. What these chapters emphasise is the link between architecture type and problem solving capability or performance.

Kerber reflects upon the diverse types of mathematical proof that are possible for a human prover and recognises the limited progress in transferring many of these proof strategies to artificial fully automated systems. He then demonstrates both formal and informal strategies that can be used to solve McCarthy's mutilated checker board problem. In a forward-looking end-section, Kerber then speculates about the future prospects for automated theorem proving. He suggests that artificial systems that rely upon using a small set of proof rules applied over formal axioms will fail to capture human-level proficiencies that, in humans, rely upon being able to think flexibly about proofs, re-represent problems and to extend the proof rules they use. So Kerber concludes that a search for a more complex and human-like architecture is likely to be required. Where Kerber focuses on discussing the details of algorithms and problem representations that can provide proofs, he recognises how the potential scope of proof solutions is constrained by the architectures upon which algorithms operate. This end-point provides a fitting link to the next two chapters by Logan and Scheutz which both provide an analysis of information processing architectures in terms of formal models, with a notion of correctness and proof for possible interactions within abstractly described architectures.

Logan starts his analysis of agent systems by posing several challenges: is it possible to formally characterise their properties in terms of an architectural description?; and is it possible to verify if two systems have equivalent architectures? He then goes on to show how an agent system can be decomposed into two components: an agent program and an architecture. The relationship between these components is that the agent program runs on the agent architecture with the architecture defining the atomic operations of the program; and also the automatic operations that will occur without the program having to do anything. In a more formal sense Logan shows the 'pure' architecture is a language for describing states and transitions and their constraints. At this level a logic can then be used to axiomatise these states and transitions, which Logan demonstrates for several agent programs. Scheutz's contribution approaches a discussion of architectural issues from a very different initial position to Logan. Scheutz is motivated to elucidate how the concept of supervenience can be used to solve the "mind-body problem". The particular approach then taken is to view supervenience in terms of a rich notion of implementation and so redefine this problem as an instance of a more general problem of implementation of virtual machines by other virtual machines or physical machines. Scheutz proceeds by considering how function units of architectures (subarchitectures) might be identified. Scheutz wields his analytic knife in splitting architectures into subarchitecture functional units organised around input, output, and inner states with transition functions between them. This brings Scheutz's contribution to how a formal specification of a virtual machine architecture can be produced. Scheutz's ultimate aim is to show how one virtual machine can be implemented in another in a non-reductive fashion. So whilst the higher level virtual machine architecture is encoded in or 'made out of' a lower level virtual machine architecture the parts of the higher level architecture are not equated with parts of the lower level architecture in a reductive fashion. Whilst there are clear commonalities between Logan and Scheutz's treatment, there are also clear differences. Scheutz is concerned with virtual machines whereas Logan's treatment applies to real or virtual machines. Logan also arrives at this discussion from an interest on agent technology rather than philosophy of mind and so makes a distinction between agent programs and agent architectures because this is how these systems are used in real-world scientific and engineering applications.

The four chapters by Hawes, Baxter, Zillich and Wyatt all have a strong focus on robotics. However, even within this robotic theme there is considerable diversity. Hawes continues with the theme of architecture focused research by proposing architectural design patterns for future roboticists. He provides a less formal definition of an architecture than that presented by Logan or Scheutz, and which is grounded in his day-to-day use of architectures in the design of intelligent systems. He then goes on to provide more details of how architectural ideas are used in practice: in software toolkits and as design patterns. He concludes by suggesting that increased use of design patterns will help resolve the problem of transfer of architectural designs from existing implementations to new projects that ends with ideas being reinvented or re-implemented. This problem arises because detailed implementations are often too specialised for wide reuse. By focusing on patterns rather than detailed implementations, Hawes suggests reuse of what is best reusable in architectural descriptions will be facilitated without the hindrance of including irrelevant system specific details.

Zillich addresses current problems and solutions in computational vision. In particular he reflects upon how the existing specialised solutions in vision might be integrated within a unified theory of computational vision. He relates vision to the whole agent problem as whole agents need to do more than just reconstruct and interpret 3D scenes. These tasks include vision for locomotion, manipulation, learning, recognition, aspects of communication and social interactions more generally. He presents limitations for current vision research in abstraction, integration and dealing with failure. Zillich then finishes by assessing the prospects for vision research, in particular vision as conceived as prediction. The topic moves from prediction in vision to prediction more generally with Wyatts contribution. In this chapter Wyatt starts with a review of work on machines that can predict the outcomes of actions on objects. This is useful for robotics, but it also raises questions about what kinds of predicting machines there can be. Wyatt suggests that some kinds of predictors fit what Sloman has called a Kantian model of prediction, whereas others fit an inductive model that is what Sloman would call Human. Wyatt speculates about why prediction is useful, how animals evolved the ability to predict, and whether there is a basic difference between the kinds of predictors that animals have in their brains, and prediction as performed using natural laws in science.

As we have described above, many of the chapters have strong links to applications. Baxter continues the robotics theme but with a stronger focus on applied robotics. The next three chapters each discuss application areas more centrally and show how progress can be made by integration of cognitive theories into applied domains. Baxter provides a historical reviews of developments in architectures for combining planning and acting in robots. He considers industrial and other real-world applications that include planning and execution systems used by NASA in their Deep Space 1 experimental mission and in Unmanned Aerial Vehicles (UAVs). He concludes by showing how the CogAff schema provided inspiration in his applied robotics design work. For example, highlighting the benefits of meta-management mechanisms in avoiding repetitive planning loops or to prevent overplanning.

Beaudoin shows how a design-based approach to understand motivation can help overcome challenges for expert progressive problem solvers who need to learn material in depth. The underpinning theory of motivation presented by Beaudoin is based upon the theory of motive processing previously developed by Sloman (1993) and Beaudoin (1994) himself. This motive processing theory is significantly different to how motivation is currently conceived in some areas of psychology but it has significant commonalities with other chapters in this collection. For example, the distinction between motives that are processed attentively in the deliberative component of the H-CogAff architecture or are processed without attention in the reactive component is at the core of the explanations for several emotional phenomena described by Petters. Beaudoin's contribution therefore links an applied area of research—productive working—with a new treatment of the concept of motive-generativators. This chapter also describes the technique of conceptual analysis and illustrates the benefits of this techniques for productive working. This is therefore another strong link to the work of Aaron Sloman, who has used conceptual analysis widely (Sloman 1978, 1982).

Khabaza presents material linking theories on cognitive science to data mining applications. He reviews observations and generalisations on data mining, presenting these as the 'Nine Laws of Data Mining'. Khabaza also speculates that future data mining systems will not only retrieve patterns in data as an outcome of the mining process but may also help produce insight across the different stages of data mining including data preparation. In addition, Khabaza reviews an aspect of Aaron Sloman's legacy not covered by any of the other chapters in this collection. This is how the

Poplog AI programming environment developed by Aaron Sloman and others at Sussex University came to be incorporated within commercial products including the Clementine data mining system. Read and Barcena present the challenges of second language learning and describe an innovative framework for use in 'Intelligent Computer Assisted Language Learning'. This system utilises four knowledge models dealing with conceptual, linguistic, collaborative and content domains. The necessary linguistic, conceptual and functional knowledge in these models is then structured and interrelated by a meta-model. Read and Barcena conclude by situating the current systems design in the context of previous design iterations and speculate about future directions.

Whereas many of the chapters in this collection are situated at the level of whole architectures of people, other animals or robots, the final chapter takes a very much more fine-grained approach. Wright reflects upon how some simple artefacts can possess intentional properties. Wright suggests that if this can be demonstrated, then it allows an understanding of how intentional properties are reducible to non-intentional properties. This is then applicable to explaining cognition in general, and hence all the natural or artificial systems presented in preceding chapters.

Earlier versions of many of the contributions in this collection were presented in a symposium at the University of Birmingham in September 2011—with Aaron Sloman present. Aaron has continued to publish in a wide variety of research areas and since the symposium Aaron has published over 15 chapters including material on: 'Biological, computational and robotic connections with Kant's theory of mathematical knowledge' (Sloman 2014); 'Meta-morphogenesis and the Creativity of Evolution' (Sloman 2012); and four chapters in a prize winning book on Alan Turing which combines chapters by Turing with newly written commentaries (Sloman 2013a, b, c, d). These four pieces cover topics including virtual machines, evolution, the Turing test, and Turing's concept of morphogenesis. In addition, at the Artificial General Intelligence 'Winter Intelligence at Oxford University in December 2012, Aaron Sloman was interviewed by Adam Ford on his views about AI, its future and its relationship to the Meta-Morphogenesis project. A video recording of the interview made available on YouTube led Dylan Holmes (MIT) to produce a complete transcript, entitled "Artificial Intelligence, Natural Intelligence and Evolution". This was checked and extended by Aaron with reference to his recent work and made freely available here: http://www.cs.bham.ac.uk/research/projects/cogaff/ agi-interview-sloman.pdf (Sloman and Ford 2014). However, it was not possible to include the transcript in this volume because it was covered by the Creative Commons Attribution license. This edited and extended version starts with Aaron describing his academic background—from mathematics to philosophy and then AI. There follow sections concerned with what problems of intelligence evolution solved, how public languages and internal states are related, and the range of possible intelligences from non-human animal intelligence to artifical general intelligence.

References

Beaudoin L (1994) Goal processing in autonomous agents. Ph.D. thesis, School of Computer Science, The University of Birmingham. Available at http://www.cs.bham.ac.uk/research/cogaff/

Sloman A (1978) The computer revolution in philosophy. Harvester Press (and Humanities Press), Sussex. http://www.cs.bham.ac.uk/research/cogaff/crp

Sloman A (1982) Towards a grammar of emotions. New Univ Q 36(3):230–238

Sloman A (1993) Prospects for AI as the general science of intelligence. In: Sloman A, Hogg D, Humphreys G, Partridge D, Ramsay A (eds) Prospects for artificial intelligence, pp 1–10. IOS Press, Amsterdam

Sloman A (2000) Architectural requirements for human-like agents both natural and artificial. (what sorts of machines can love?). In: Dautenhahn K (ed) Human cognition and social agent technology. Advances in consciousness research, pp 163–195. John Benjamins, Amsterdam

Sloman A (2012) Meta-morphogenesis and the creativity of evolution. In: Besold T, Kuehnberger K, Schlorlemmer M, Smaill A (eds) proceedings of ECAI 2012 workshop on computational creativity, concept invention, and general intelligence, pp 48–55. ECAI, Montpellier

Sloman A (2013a) Aaron sloman develops a distinctive view of virtual machinery and evolution of mind (part 1). In: Cooper S, van Leeuwen J (eds) Alan Turing—his work and impact. Elsevier Science, Amsterdam, pp 97–101

Sloman A (2013b) Aaron sloman draws together—absolves alan turing of—the mythical turing test. In: Cooper S, van Leeuwen J (eds) Alan Turing—his work and impact, pp 606–610. Elsevier Science, Amsterdam

Sloman A (2013c) Aaron sloman draws together—virtual machinery and evolution of mind (part 2). In: Cooper S, van Leeuwen J (eds) Alan Turing—his work and impact, pp 574–579. Elsevier Science, Amsterdam

Sloman A (2013d) Aaron sloman travels forward to—virtual machinery and evolution of mind (part 3). Meta-morphogenesis: evolution of information-processing machinery. In: Cooper S, van Leeuwen J (eds) Alan Turing—his work and impact, pp 849–856. Elsevier Science, Amsterdam

Sloman A (2014) Biological, computational and robotic connections with kant's theory of mathematical knowledge. AI commun 27(1):53–62, (an invited paper for the"Turing and Anniversary Session" at ECAI2012)

Sloman A, Ford A (2014) Interview: artificial intelligence, natural intelligence and evolution. http://www.cs.bham.ac.uk/research/projects/cogaff/agi-interview-sloman.pdf. Accessed 26 Feb 2014

Turing A (1950) Computing machinery and intelligence. Mind 59:433–460, (reprinted in Feigenbaum EA, Feldman J (eds) Computers and thought, pp 11–35. McGraw-Hill, New York, 1963)

Wright I, Sloman A, Beaudoin L (1996) Towards a design-based analysis of emotional episodes. Philos Psychiatry Psychol 3(2):101–126

Chapter 2
Aaron Sloman: A Bright Tile in AI's Mosaic

Margaret A. Boden

2.1 The Aims of AI

When AI was still a glimmer in Alan Turing's eye, and when (soon afterwards) it was the new kid on the block at MIT and elsewhere, it wasn't regarded primarily as a source of technical gizmos for public use or commercial exploitation (Boden 2006, p. 10.i-ii). To the contrary, it was aimed at illuminating the powers of the human mind.

That's very clear from Turing's mid-century paper in *Mind* (1950), which was in effect a manifesto for a future AI. Like Allen Newell and Herbert Simon too, whose ground-breaking General Problem Solver was introduced as a simulation of "human thought" (Newell and Simon 1961). Turing's strikingly prescient plans for a wide-ranging future AI were driven by his deep curiosity about psychology. Indeed, most of AI's technical problems and answers arose in trying to discover *just how* computers could be programmed to model human thought.

Virtually all of the philosophers who read Turing's *Mind* paper ignored that aspect of it (Boden 2006, p. 16.ii.a-b). They weren't interested in the technical or psychological questions. Instead, they focussed on criticizing the so-called Turing Test–which had been included by the author not as serious philosophical argument but as jokey "propaganda" (Gandy 1996, p. 125). In other words, they responded not by getting intrigued, or even critically sceptical, about a potential AI, but by considering one ancient philosophical question: whether a machine could think.

While ignoring what Turing saw as the most important part of the paper, however, they were addressing the second main aim of early AI. For the AI pioneers weren't targeting only psychology: they had their guns trained on philosophy, too. In other words, besides wondering how the mind-brain actually works (and how it could be

M. A. Boden(✉)
University of Sussex, Falmer,Sussex, UK
e-mail: m.a.boden@sussex.ac.uk

J. L. Wyatt et al. (eds.), *From Animals to Robots and Back: Reflections on Hard Problems in the Study of Cognition*, Cognitive Systems Monographs 22, DOI: 10.1007/978-3-319-06614-1_2, © Springer International Publishing Switzerland 2014

modelled in a computer), they wondered how it is possible for a material system to have psychological properties at all.

Turing himself was interested in the philosophical problem of the nature of mind, even though he rejected the philosophers' usual way of addressing it. As for his AI successors, they hoped to illuminate a variety of long-standing problems—not only in the philosophy of mind, but in philosophical logic and epistemology as well. Minsky's (1965) paper on self-models and freewill, and the foray into logic and epistemology by McCarthy and Hayes (1969), were early examples. So was Allen Newell and Herbert Simon's approach to the nature of reference, choice, and freedom (Newell 1973). (My own 1960s work, likewise, used AI to address problematic issues in philosophy, as also in psychology: Boden (1965, 1969, 1970, 1972). Before then, I had approached these issues in a more traditional manner: Boden 1959.)

Today, AI looks very different. Most current workers avoid psychological modelling, and aren't drawn to philosophy either. Trained as computer scientists at university, they have scant interest in these matters. Originally, everyone came into AI from some other discipline, carrying a wide range of interests with them. But that's no longer so. Indeed, AI has adopted a third aim: most AI research is now directed at the useful gizmos.

(There are some exceptions, of course. Besides individual researchers—including both long-time AI modellers and younger workers—there have been some national research programmes focussed on science rather than gizmos. For example, since 2003 the European Union has provided significant funding for interdisciplinary projects on cognition, involving psychologists, philosophers and neuroscientists as well as AI programmers. These include the flagship Human Brain Project and funding devoted to ICT and creativity.)

As for the gizmos, there are lots of those. In fact, this research has been so extraordinarily successful that its results are now largely taken for granted. In other words, AI has become well-nigh invisible to the general public, who benefit from the gizmos every day in countless ways but don't realize that they utilize AI. Even computer professionals, who should know better, sometimes deny the involvement of AI: an AI friend of mine was shocked to be told by a colleague that "Natural language processing (NLP) isn't AI: it's computer science". Partly because of this invisibility, AI is often even said to have "failed" (Boden 2006, p. 13.vii.b).

There's another reason, however, why AI is commonly said to have failed: its early hopes—and hype—about modelling, matching, and maybe even surpassing human mental powers have not been met. Most of the AI gizmos (which started with work on expert systems: Boden 2006: pp. 10.iv.c, 13.ii.b) address only a very narrowly defined task, although admittedly they frequently achieve far better results than people can. Sometimes, they can surpass the world champion—as IBM's Deep Blue did when playing chess against Gary Kasparov in 1997. But even the more widely aimed AI systems don't match up to human intelligence.

Despite amazing advances in NLP, for instance (using powerful statistical methods very different from traditional AI), the ability of computers to deal with natural language is far less sensitive than that of reasonably articulate human beings. Commonsense reasoning, too, despite the power of today's data-mining and the notorious mastery of IBM's Watson in playing *Jeopardy* (Baker 2011), has not been reliably emulated. As for the non-cognitive aspects of mind, namely motivation and emotion, these are commonly ignored by AI researchers—or addressed in a very shallow fashion. In short, all these aspects of human mentality are much more challenging than was initially expected.

Even vision, a sense we share with many other animals, has turned out to be more difficult than expected. In 1966, Minsky asked a bright first-year undergraduate (Gerald Sussman) to spend the summer linking a camera to a computer and getting the computer to describe what it saw. (They wanted a vision system for the MIT robot.) This wasn't a joke: both Minsky and Sussman expected the project to succeed (Crevier 1993, p. 88). Even in the world of shadowless convex polyhedra, that was a tall order: it couldn't be mastered in a single summer. In more realistic worlds, beset not only by shadows but also by curves, multiple occlusions, and missing and/or spurious parts, visual computation is hugely more complex than had been thought. (It's still the case that computer systems presented with a complex visual scene are very limited in the objects and relationships that they can identify; moreover, they mostly ignore the biological functions of vision—see below.)

2.2 AI as Philosophy

What does all this have to do with Aaron Sloman? Well, he has never been a gizmo-chaser. With a background in mathematics and philosophy, he has always taken both the aims of early AI seriously. That was already evident in his book *The Computer Revolution in Philosophy* (1978), and remained so in his later papers. (His book is now available online, and is constantly updated—so can act as an extra reference for most of the topics mentioned here.)

But for him, the usual intellectual priorities of AI, where psychological and technical questions took precedence over philosophical ones were equalized—or even reversed. In other words, what I've called AI's "second" aim has been Sloman's first. Despite being a highly accomplished AI programmer, and a key figure in the development of AI in the UK, Sloman still regard himself as first and foremost a philosopher. ("Still", because he was a professional philosopher for some years before encountering AI.)

The familiar philosophical puzzles he has written about include freewill, reference, intentionality, representation, modal reasoning, philosophy of mathematics, causation and consciousness (Sloman 1974, 1986, 1987a, b, 1996b, c; Sloman and Chrisley 2003). His book has two illuminating chapters on the philosophy of science. Above all, however, he has been concerned with the philosophy of mind—not only the age-old problem of mind and machine (e.g. Sloman 1992, 1993), but also deep questions about the nature of mind that are *not* familiar philosophical chestnuts, as those just listed are.

One of the less familiar puzzles he has addressed is the nature of non-logical or "intuitive" thinking, such as mathematical reasoning based on diagrams. He was drawn to this topic partly because early AI wasn't able to model it. But his motives weren't purely technological. Although the paper he published on it appeared in the *Artificial Intelligence* journal, and was later included in a high-profile AI collection on knowledge representation (see below), Sloman chose to describe it in the title as exploring "Interactions Between Philosophy [*sic*] and AI" (1971: 209). Much of the paper contrasted non-logical ("analogical") reasoning with the type of inference described by the philosopher Gottlob Frege. But the main inspiration for the paper was the philosopher Immanuel Kant. Sloman wanted to defend Kant's (unfashionable) claim that intuitive mathematical reasoning, neither empirical nor analytic, could lead to necessary truths—a defence that he'd mounted at greater length in his DPhil of 1962.

Sloman has never been guilty of irresponsible hype about AI, which he has criticized consistently over the years. For instance, in his 1978 book (Sect. 9.12) he predicted that , by the end of the century, computer vision would not be adequate for the design of domestic robots capable of washing dishes, cleaning up spilt milk, etc. Nevertheless, he has always had high ambitions for the field. Avoiding the narrow alleyways of gizmo-AI, his prime concern has been the mind as a whole.

Or perhaps one should rather say *minds* as a whole, since he has considered intelligence in general—in animals, humans and machines. Besides remarking on the mental architecture of particular species (e.g. humans, chimps, crows, ants...), he has tried to outline the space of all possible minds (Sloman 1978, Chap. 6). His work makes it clear that intelligence isn't an all-or-none phenomenon, nor even a continuously varying property. Rather, it's a richly structured space defined by many distinct dimensions, or information-processing procedures, which enable radically different—and to some extent incommensurable—types of intelligence to arise.

To some extent, Sloman's view of mind leant on the philosophy of Ryle (1949), whom he had encountered as a DPhil student at Oxford (Sloman 1996c: Acknowledgments; 1978: Chap. 4). Besides always sharing Ryle's scorn for "the ghost in the machine", Sloman was—eventually (see below)—inspired by his talk of "dispositions" versus "episodes".

Ryle analysed many psychological terms not as reports of actual events or phenomena, but as denoting long-standing dispositions to behave in a certain way in certain circumstances. He compared jealousy, for example, with brittleness. To say that glass is brittle is to say that *if* it is hit *then* it will probably break; likewise, to say that someone is jealous is to say (for example) that *if* someone sees her husband talking animatedly to another woman *then* she will very likely say something unpleasant to one or both of them. The same is true, he argued, of concepts denoting propositional attitudes—such as *know, believe, desire, prefer, fear, expect*, and *hope*. So to believe that *p* is to be disposed to say that *p*, and to behave in ways that would be appropriate (given the person's other beliefs and desires) if *p* were true.

This approach implied that most psychological concepts are logically interlinked with others. In other words, the activation of disposition *a* is likely (by definition) to lead to the activation of dispositions *b, c, ...* and to the triggering of episodes *x, y*, and *z*. To be sure, dispositions can be suppressed—much as a piece of glass may never be dropped, or may be wrapped in a protective cloth to prevent its breaking.

But jealousy without *any* tendency to resent, denigrate, or harm the person or persons concerned simply is not jealousy.

Even first-person psychological statements such as *I see blue* or *I have an itch*, said Ryle, are not reports of events in some mysterious non-material world but "avowals" of certain behavioural dispositions. So someone who claimed to see blue, or to have an itch, who did not assert any resemblance with the sky, or make any attempt to scratch, would either be thought insincere or would simply not be understood. As for the feelings sometimes involved in emotions, to say *I feel depressed*, according to Ryle, is not to report an internally accessible conscious state, but rather to perform "a piece of conversational moping: … not discovery [by Cartesian direct access], but voluntary non-concealment" (Ryle 1949, p. 102).

Ryle was widely accused of behaviourism—and, in my view, rightly so. However, his key term *disposition* was systematically ambiguous, denoting either observable behaviour and/or its underlying causes. Most analytic philosophers in the 1950s, like Ryle himself, interpreted it as a summary *description* of behaviour, not an *explanation* of it. This was largely because they saw explanation as the task of science, not philosophy (see Sect. 2.5). Later, the term was read by some (not all) philosophers as explanatory, denoting *the mechanism responsible for the relevant behaviour* (Squires 1970).

Initially, Sloman's reading of Ryle's key term was descriptive rather than explanatory. As a result, he rejected Ryle as a behaviourist. Moreover, if he'd been asked to interpret "disposition" as explanation, he would have assumed it to refer to some (unknown) neural mechanism. But his conversion to AI enabled him to see that it could also be a *computational* explanation.

On re-reading *The Concept of Mind* (Ryle 1949), he was especially interested by the fact that—as remarked above—Ryle's dispositions and episodes were interlinked. In other words, one mental state could switch to another mental state, much as one part of a computer program could activate another. His work thereafter can be seen as an attempt to put specific computational flesh onto broadly Rylean dispositional bones.

2.3   A Vision of Vision

Sloman's avoidance of AI hype is grounded in his nuanced appreciation of the significant complexity and diversity of human (and much animal) intelligence. That was apparent even in his earliest work on computer vision, the POPEYE project (Sloman 1978, Chap. 9).

POPEYE modelled the interpretation not of fully connected drawings of perfect polyhedra (or even polyhedra-with-shadows), but of highly ambiguous, noisy, input—with both missing and spurious parts: see Fig. 2.1. And it simulated the complexity of perception to an extent that was highly unusual at the time.

Sloman's thinking about vision (although not the POPEYE program itself), especially in the years following the implementation of POPEYE, stressed the fact that

Fig. 2.1 Example of the kind of ambiguous stimuli that POPEYE operated upon. When viewed together the ambiguous fragments of this picture can be recognized as letters forming a familiar word

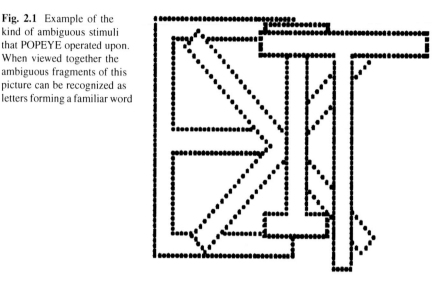

vision is integrated with action and motivation (Sloman 1983, 1989). This was a lesson that he had learnt from the psychologist Gibson (1966). Gibson's theory of perceptual "affordances" held that vision has evolved for a range of different purposes, for which different types of motor action are appropriate.

That is, the primary point of vision is not to build a visual image, nor even—as David Marr would argue later (see Sect. 2.5)—to represent the location of objects in 3D-space. Rather, it is to prepare for and guide motor behaviour, enabling the organism to achieve evolutionarily significant goals. As well as answering questions about what things are in the environment and just where they are located, such goals include recognizing and following a pathway, avoiding an obstacle, jumping onto a stable support, approaching a potential mate and deciding which way to move in order to see more of something already glimpsed.

If those are the purposes of vision, a realistic (or even near-realistic) computer model would need to combine many different sorts of background knowledge. Moreover, information processing could occur concurrently in different domains, determining which subprocesses would dominate the scarce computational resources. (This differed from the heterarchy so popular in the early-mid 1970s, wherein there was only one locus of control at any moment, and control was transferred to process X by an explicit call from process Y: Boden 2006, p. 778ff.). Each knowledge domain in POPEYE had its own priorities for finding/processing information, and these priorities could change suddenly as a result of new information arriving unexpectedly. Diversity (and flexibility) was increased also by the fact that some of the internal representations constructed by POPEYE were temporary, rather than provisional. (Something provisional may become permanent, but something temporary should not.)

Considered as a practical visual system for a robot, POPEYE wasn't impressive. Quite apart from anything else, it didn't actually contain anything that linked to bodily action (though some aspects of it could have been so linked, if Sloman had had the opportunity to develop it further: see Sect. 2.5). But Sloman wasn't trying to advance robotics, least of all robotics confined to toy polyhedral worlds. Rather, he was trying—again, a *philosophical* aim—to advance Kant's argument that the mind must provide some prior knowledge for even the "simplest" perceptions to be possible (Sloman 1978, p. 230). But whereas for Kant the principles of organization were very general, and innate, for Sloman they also included highly specific learnt examples.

Accordingly, his program modelled the fact that high-level visual schemata can aid recognition enormously. For instance, learnt knowledge of the familiar upper-case sign "EXIT" helps us—and POPEYE—to recognize the four letters in Fig. 2.1. This computational diversity has a chicken-and-egg aspect: if one recognizes a particular set of dots in Fig. 2.1 as co-linear, that can help one to recognize an "E"; but if one has already recognized "EXIT", one will be much more likely to recognize *those very dots* as co-linear (Sloman 1978, pp. 228–232).

The moral of POPEYE, as of Sloman's more recent work on vision (1983, 1989), was that any realistic degree of visual complexity will involve many diverse types of background knowledge, all playing their parts concurrently. As he put it: "Our program uses knowledge about many different kinds of objects and relationships, and runs several different sorts of processes in parallel, so that 'high-level' processes and (relatively) 'low-level' processes can help one another resolve ambiguities and reduce the amount of searching for consistent interpretations. It is also possible to suspend processes which are no longer useful: for example low-level analysis processes, looking for evidence of lines, may be terminated prematurely if some higher level process has decided that enough has been learnt about the image to generate a useful interpretation. This corresponds to the fact that we may recognize a whole (e.g. a word) without taking in all of its parts" (1978, p. 229).

In an important sense, however, POPEYE wasn't really—or anyway, it wasn't *only*—about vision. Rather, it was a preliminary exercise in architecture building. For Sloman saw computer vision as a way of keying in to the computational structure of the mind as a whole.

2.4 Architectural Issues

In his early book, Sloman had discussed the nature of the mind as a whole (1978, Chap. 6). He argued, for example, that because emotion is integral to intelligence, truly intelligent robots would have to have emotions too. For instance "they will sometimes have to feel the need for great urgency when things are going wrong and something has to be done about it" (1978, p. 272; cf. Sloman and Croucher 1981; Sloman 1982).

Over the following years, he focussed increasingly on the control structure of the entire mind, eventually offering computational analyses of motivation and emotion that illuminated even such seemingly computationally recalcitrant phenomena as anxiety and grief. (The most accessible statement of his mature approach is Sloman 2000; for more technical descriptions, see Sloman 1998, 2001, 2003, and Sloman 2014)

Emotions often involve conscious feelings, but—Sloman argued—these are not all there is to emotion. Emotions are control-structures, participating in the guidance of action and the scheduling of potentially conflicting motives. They enable goals and subgoals to be chosen appropriately, and—if necessary—to be put on hold, or even dropped, as circumstances change. They interact with perception, and with various types of short-term and long-term memory, alarm systems and (variable) attention-thresholds.

According to Sloman (and his student Luc Beaudoin), emotions—and whole minds, too—differ from each other in terms of three main architectural levels. These involve what he calls reactive, deliberative and meta-management mechanisms.

The minds of insects are mostly reactive, depending on learnt or innate reflexes. They are capable only of "proto-emotions": inflexible reactions that have much the same adaptive function as (for instance) fear in higher animals.

A chimpanzee's mind is largely deliberative, capable of representing and comparing past, and possible future, actions or events. So the animal is capable of backward-looking and forward-looking emotions: non-linguistic versions of anxiety and hope, for example.

In general, deliberative mechanisms are more complex than reactive ones. They are also diverse, including various intermediate levels between pure reaction and full-blown planning—which employs multi-step look-ahead with various strategies, and uses meta-management (the third level) to control the planning process (Sloman 2009a). (This diversity is underplayed by the currently fashionable embodiment/enactive movement: e.g. Brooks 1990; Clark 2013. Among other things, such work prioritizes instant control via evanescent "online" signals, at the expense of more long-lasting "offline" processes and data-structures: Sloman 2009b, 2013.)

In human adults, the deliberations can include conscious planning and reasoning, generating more precisely directed emotions accordingly. In general, language makes possible emotions with propositional content, which may be highly specific—and which may vary significantly from one culture to another. The concept of love, for example, differs across cultures: so emotions such as love and grief, and even honour, differ too. In addition, an adult human mind has a rich store of reflexive meta-management mechanisms, which monitor and guide behaviour. Emotions centred on the concept of self—such as vainglory and embarrassment—are now possible, accordingly.

Sometimes, humans seemingly have no choice: a danger that's just been identified *must* be averted, and it must be done *now*. There's no time for conscious deliberations. Reactive mechanisms must take control. But the sense in which a human being (sometimes) has no choice about what to do next is fundamentally different from the sense in which an insect (always) has no choice. Humans are free, whereas insects aren't. But human freedom doesn't depend on randomness, or on mysterious spiritual influences: to the contrary, it's an aspect of *how our minds work.* Sloman's account of mental architecture shows how our emotions can sometimes compromise our freedom (by leading us to react [*sic*] unthinkingly) even though they also help to make it possible (by controlling appropriate cognitive mechanisms, such as deliberation).

The specific emotions discussed by Sloman include grief and sorrow, closely related but different emotions that are generated by the death of a loved one. A dog may suffer from sorrow, and appear to mourn its lost master. But grief in a human mourner is a much more complex, and ("Edinburgh Bobby" notwithstanding) more long-lasting, than it is in a dog. Quite apart from being expressible in a host of linguistically distinguishable ways, it leads to continual (though gradually decreasing) interruptions of thinking and behaviour as the mourner remembers, or is reminded of, the lost person. The previously built motivational structure of caring about and encouraging the goals and interests *of the loved person* has to be gradually dismantled (Fisher 1990). This is not, and cannot be, the work of a minute: mourning inevitably takes time.

Most of what was said in the previous paragraph could have been said by Ryle— or by a competent novelist. But in discussing grief, Sloman and his students (Beaudoin and Ian Wright) used their deep knowledge of AI to suggest a host of specific information-processing mechanisms that could interact to generate the various mental/behavioural phenomena concerned (Wright et al. 1996; Sloman 2000).

Critics will surely complain that grief, over and above its dispositional aspects, involves searing feelings, conscious episodes which—they say—cannot be captured in computational terms. Indeed, many say this about emotions in general. Feelings of grief, or joy, or anxiety … are special cases of what philosophers call *qualia*. Any adequate theory of emotion must therefore make place for qualia. But this is a tall order. For, notoriously, *all* philosophers (Rene Descartes and Ryle included) have difficulty in giving a coherent account of conscious feelings or sensations.

In other words, the AI-friendly thinkers aren't the only ones to encounter trouble here. But, undeniably, trouble there is. Some computationalists have denied the existence of qualia (Dennett 1988, 1991, Chap. 12). Sloman did not. Instead, he analysed them as aspects of the virtual machine which is the mind (1999; Sloman and Chrisley 2003).

Specifically, he saw them as intermediate structures and processes generated by an information-processing system with a complex, reflexive, structure. Some qualia (but not all) can be noticed and thought about using self-reports—which might require the system to generate ways of classifying them, using *internal* categories that can't be matched/compared with comparable categories in other virtual machines (other minds). But the self-reports are something extra. They are directly accessible to the highest level of the system itself, and are sometimes communicated verbally, or expressed behaviourally, to others.

(On this view, some of the very same qualia could exist in simpler organisms that have sophisticated perceptual mechanisms without also having human-like self-monitoring mechanisms for introspection. So a house-fly might have visual qualia of which it simply cannot be aware. Clearly, Sloman's analysis conflicts with any view which requires qualia, by definition, to involve self-knowledge, or to be actually attended to.)

As Ryle would doubtless have been glad to hear, self-reported qualia do not rest on Cartesian "direct access" to some mysterious mental world. The directness, or lack of evidence, of first-person experiential statements is due—so Sloman argues— to the particular kind of (reflexive) computation involved. For example, the meta-management system may have access to some intermediate perceptual data-base (*blue, itch* ...) which does not represent anything in the third-person-observable outside world because it is the content of a dream or hallucination. In other cases, it would be part of the process of perceiving something external.

Sloman's pioneering discussions of the integration of cognition, motivation, and emotion were *computational* analyses, in the sense that they conceptualized the mind as an information-processing system and were deeply informed by an extensive knowledge of various types of AI research. And they have been extended (and are still being developed) by him and his colleagues in the same strongly computational spirit. But, for many years, they were not illustrated by functioning computer programs. Even now, Sloman cannot provide a computer model of his architectural theory as a whole.

Since the 1990s, however, he and his students have implemented a model based on his theory of emotional perturbances. This is the series of MINDER programs, developed to illuminate the nature of emotion, and its role in the control of action (Wright and Sloman 1997; see also Beaudoin 1994; Wright 1997). (To *illuminate*, not to capture: these programs model only a very limited subset of Sloman's theory of the mind.)

MINDER simulates the anxiety that arises within a nursemaid, left to look after several babies single-handed. She has only a few tasks: to feed them, to try to prevent them from falling into ditches, and to take them to a first-aid station if they do. And she has only a few motives to follow: feeding a baby; putting a baby behind a protective fence, if one already exists; moving a baby out of a ditch for first-aid; patrolling the ditch; building a fence; moving a baby to a safe distance away from the ditch; and, if no other motive is currently activated, wandering around the nursery. In short, she's hugely simpler than a real nursemaid. Nevertheless, she is prone to emotional perturbations ("proto-emotions") comparable to anxiety—or rather, to several interestingly different types of anxiety.

Sloman's simulated nursemaid has to notice, and respond appropriately to, a number of visual signals from her environment. Some of these trigger (or affect) goals that are more urgent than others: a baby crawling towards the ditch needs her attention sooner than a merely hungry baby does, and one who's about to topple over the edge of the ditch needs attention sooner still. But even those goals which can be put on hold for a while may have to be coped with eventually; and their degree of urgency may rise with time. So a near-starving baby, who has not been fed for hours, can be put back into its cot if another baby is about to fall in the ditch; but the baby who has waited longest to be fed should be nurtured before the ones whose last meal is more recent. As these examples suggest, the nursemaid's various tasks can be interrupted, and either abandoned or put on hold. She—or rather, the MINDER program—has to decide just what the current priorities are. Much as with a real nursemaid, her anxieties increase, and her performance degrades, with an increase in the number of babies—each of which is an unpredictable autonomous system.

Sloman has identified several important limitations of MINDER (Wright and Sloman 1997: Sects. 3.7.2, 4.3), some of which could be alleviated or overcome in later versions. But it's important to note that this system could be used to model many different types of autonomous agent besides babies and nursemaids. Indeed, a programming environment based on the early work on MINDER (and first used in Wright 1997) has been placed on the Internet for other AI workers to experiment with. This is the SimAgent toolkit (Sloman and Poli 1995; Sloman 1995), which has been used by a number of researchers outside Sloman's group.

We're now in the twenty-first Century, with POPEYE and MINDER approaching the status of ancestral forms. Their descendants in Sloman's recent thinking include his current work on autism (soon to be included on his website).

This work-in-progress attempts to throw light on the intriguing drawing-abilities of the autistic child Nadia, and to explain why they regressed when her language skills developed. It also suggests that autism considered as a deficiency in Theory of Mind (Frith 1989/2003; Boden 2006, p. 7.vi.f-g) is a special case of a more general range of possible developmental abnormalities that can impede later developments. Sloman situates these ideas within the theoretical framework that he (with the ethologist Jackie Chappell) has produced for accounting for "the differences between precocial and altricial species, where the latter have multiple routes leading from genome to behaviours, through competences that develop late and build on competences that developed earlier under the influence of the environment" (p.c.).

Autism isn't the only intriguing topic that Sloman is currently working on. Meta-morphogenesis is another. This is an enquiry into how the possible "design spaces" and "niche spaces" in biological evolution generate, and are generated by, an increasing variety of information-processing mechanisms. How is morphogenesis driven by new forms of representation, ontologies and architectures? How do reflexive (meta-management) mechanisms evolve that can self-monitor and self-modify the developing organism? And can special-purpose chemical mechanisms, which involve both continuous and discrete changes, produce results that cannot be produced, or cannot be produced quickly, by Turing-computation?

A third current concern is a development of the Kantian position that he initiated as a graduate student in mathematics, before switching to philosophy. As he puts it: "I've been exploring the conjecture that the precursors of Euclid must have started from something like more general versions of Gibson's abilities to perceive and reason about positive and negative affordances. I've also been trying to isolate the specific forms of human mathematical (especially geometrical) reasoning that current forms of logical, arithmetical and algebraic theorem proving in AI fail to capture" (p.c.). For example, one can *see* that the angles of a triangle *necessarily* continue to add up to the 'angle' of a straight line as the size and shape of a triangle change, because one has intuitive (non-logical) knowledge of the possibilities involved in the spatial structures concerned. So, again, an intriguing—and potentially game-changing—coupling of philosophy and AI.

Clearly, Sloman's intellectual curiosity, and insightfulness, is as wide-ranging and as sparkling as ever. In short, he's still a bright tile in AI's mosaic.—So:*Watch this space!* More to the point: *Watch his website!*

2.5 Recognition Delayed

There's a puzzle here, however: Sloman's work was under-appreciated for many years. The reason, in a nutshell, is that it didn't (yet) fit in with the *Zeitgeist.*

To be sure, he was always highly respected by the UK's AI community, who could interact with him personally. In verbal discussions, the air would fizz with his searching questions and unexpected insights. The AI-pioneers in the USA knew him personally too, and respected his contributions accordingly. His account of analogical representation (Sloman 1971, 1975), for instance, attracted their interest immediately, and was recognized by them as offering key—albeit "controversial"—ideas in knowledge representation (Brachman and Levesque 1985, p. 431). But the subsequent generations of USA's AI scientists were less familiar with his research.

In part, that was due to AI's no longer being a tiny research community. In addition, most of his publications were on philosophical topics. (Even these were relatively few in number, since he devoted so much time to helping other people's learning and research: see Sect. 2.7.) But the lack of recognition was due also to the fact that POPEYE and MINDER, and the integrative computational philosophy that underlay them, were deeply unfashionable.

Sloman's work on vision was already non-mainstream in the mid-1970s, as we've seen. Most computer vision at that time was conceptualized much more narrowly. But bad—or rather, unlucky—timing soon played a part too.

When David Marr's *Vision* (1982) was published, shortly after his tragically early death, it was instantly influential. Indeed, many people interested in computer vision hadn't waited for the book, having been converted by Marr in the mid-late 1970s. Marr's vision papers were readily available from MIT as AI Memos, one had been published by *Science,* and three by the Royal Society (Marr 1974a, b, 1975a, b, c; Marr and Hildreth 1980; Marr and Nishihara 1978; Marr and Poggio 1976, 1977, 1979). Quite apart from their interest as examples of AI, these publications promised to illuminate the relevant areas of neuroscience–as Marr's earlier work had done for the cortex and cerebellum (Boden 2006, p. 14.v.b-f). As a result of the huge interest that they aroused, AI scientists asking questions about vision in the late-1970s and 1980s tended to base them in Marr's approach.

In other words, they focussed almost entirely on optically specifiable (and probably innate) bottom-up processes, not on top-down influences from learnt high-level schemata. They ignored questions about the use of vision in motor control. They accepted Marr's key claim that the purpose of vision is to turn the retinal 2D-representation into a representation of the 3D-environment. They emphasized the fact—which Sloman, like most early AI-vision researchers, hadn't stressed (but see 1978, p. 219)—that we can locate and describe visible objects that we have never seen before. And they interpreted the visual scene in terms of surfaces, edges, textures and 3D-locations—not in terms of identifiable objects, and still less in terms of those objects' potential roles in the organism's behaviour.

As part of the Marrian revolution, Gibson's approach to vision—which had influenced Sloman deeply, as we've seen—was scornfully rejected. Admittedly, this was primarily because Gibson had claimed that low-level vision doesn't involve computation: inevitably, a red rag to all AI bulls (Boden 2006, p. 7.v.e-f). Marr's theory, of course, identified and modelled many detailed computations going on in low-level vision. Admittedly too, no Marrian would have denied that vision is useful, and often essential, for action. But in their discussions about and modelling of visual perception, the Marrians said nothing about how it is integrated with motor control, or with a changing motivational context. One might almost characterize their approach as the study of *vision without mental architecture*.

In all these ways, then, Marr's work was at odds with Sloman's. As a corollary to becoming deeply unfashionable virtually overnight, Sloman lost his research funding. POPEYE was now so far off the mainstream that it simply stood no chance.

(His account of what happened is given in the historical note added to the online version of his 1978 book. One of the factors he mentions is yet another aspect of the *Zeitgeist*: the widespread move from AI languages such as LISP or POP-11 to more general, more efficient, languages such as Pascal or C/C++. These, he said, make it very difficult to express "complex operations involving structural descriptions, pattern matching and searching", and to permit "task-specific syntactic extensions ... which allow the features of different problems to be expressed in different formalisms within the same larger program". Today, he says that the same rejection of AI languages may have prevented a wider take-up of the SimAgent toolkit, based as it is on the POP-11 language.)

Sloman's pioneering work on mental architecture, too, was largely ignored by AI scientists for a while. No doubt, that can be explained in part by the unfashionableness of the topic at the time. Only a few computer modellers were thinking about the mind as a whole (e.g. Simon 1962, 1969; Anderson 1983; Minsky 1985; Laird et al. 1987). And, despite Simon's having written about emotions as long ago as the 1960s (Simon 1967), even they were looking at cognition (perception, planning, problem-solving) rather than emotion and/or motivation. Within the psychological and neurological literature too, the emotional aspects of thinking were still mostly ignored. (Minsky was an exception here: his work on "the emotion machine" was circulated for several years before finally being published in 2007.) It wasn't until the 1990s that the concept of "emotional intelligence" became popular, entering not only the newspapers but also AI research (Damasio 1994; Picard 1997, 1999).

But there was an additional problem. The neglect, especially in the USA's AI community, was due also to the younger generations' dismissal of anything that wasn't actually implemented. As remarked in Sect. 2.4, Sloman's studies of architecture, including his account of motivation and emotion, is deeply informed by AI and computational thinking without being presented as computer programs. Now, to be sure, there is MINDER—and the SimAgent environment, too. But for many years no such implementation existed. And even these model only a very small part of Sloman's theory.

This shouldn't matter: although functioning programs are of course a huge strength, computational *thinking* about intelligence is valuable also. Indeed, the latter must precede the former. Both MINDER and SimAgent , after all, resulted from Sloman's methodology of developing and testing *theoretical* architectural ideas about functional requirements, evolutionary origins, and variants in other species. Nevertheless, in the quest for comprehensive programmed models, he couldn't deliver.

As for the philosophers, their professional *Zeitgeist*, also, prevented them from recognizing Sloman's importance quickly. When AI was still young they knew little or nothing about it. Most got no further than the Turing Test, even though Turing himself had made other philosophically interesting claims in his notorious *Mind* article (Boden 2006, p. 16.ii.b). Nor were many of them persuaded to learn about it when Sloman published *The Computer Revolution in Philosophy* in 1978. Besides being informed by a detailed knowledge of AI which they lacked, and which they therefore could not understand, the book boldly announced that "within a few years philosophers ... will be professionally incompetent if they are not well-informed about these developments [in computing and AI]" (1978, p. xiii). That announcement was correct—but perhaps hardly tactful. It was bound to raise many philosophers' hackles.

The philosophers' resistance wasn't based purely in annoyance at being told that they were ignorant. It was underpinned by the fact that most of them believed science to be in principle irrelevant to philosophy. (Hence their reluctance to interpret Ryle's philosophy of mind as offering *explanations:* see Sect. 2.2.) Anyone who disagreed was accused of "scientism" and/or "psychologism", which Frege—and his translator John Austin, then the high-priest of Oxford philosophy—had denounced as a near-deadly sin (Frege 1884/1950, Preface). My own first book (Boden 1972), which used both AI and psychology to inform philosophical argumentation about mind and personality, had also suffered from this attitude: one highly complimentary review in a philosophical journal ended by saying "... but you can't really call it philosophy".

Even those philosophers who, in the decade following publication of Sloman's book, became willing to grant that AI could be philosophically interesting often missed the point. For instance, when I was putting together a collection of papers on *The Philosophy of Artificial Intelligence* for Oxford University Press in the late-1980s (Boden 1990), one of the publisher's advisers said that Sloman's 'Motives, Mechanisms, and Emotions' (1987a) should be dropped. He—or (very unlikely) she—announced that it was "unrepresentative" and "irrelevant". The adviser was half-right. It was indeed unrepresentative, for it was years ahead of its time (see Sect. 2.4). But "irrelevant"...? The mind boggles. Only a narrowly technological, gizmo-seeking, view could have justified such a judgment. I insisted that the paper be included.

2.6 The Zeitgeist Shifts

However, things change. The change of most relevance here is not in Sloman's theoretical approach, although this has of course developed over the years—and is still doing so. Rather, it is in the surrounding intellectual atmosphere. The *Zeitgeist* of AI, in particular, has altered significantly.

POPEYE—or, more accurately, the general philosophy that underlay POPEYE—has recently had something of a revival. For Sloman's 1989 paper on vision was cited by the neuroscientist who, with a psychologist colleague, had recently caused a sensation by positing two visual pathways in the brain–one for perception, the other for action (Goodale and Humphrey 1998, p. 201).

According to Melvyn Goodale and David Milner, the dorsal pathway locates an object in space relative to the viewer, who can then grasp it; the ventral pathway may (this point is contested) locate it relative to other objects, and enables the viewer to recognize it (Goodale and Milner 1992, 2003; Milner and Goodale 1993). (The evidence lies partly in brain-scanning experiments with normal people, and partly in clinical cases: damage to these brain areas leads to visual ataxia and visual agnosia, respectively. For example, one patient can recognize an envelope but is unable to post it through a slot, whereas another can post it efficiently but can't say what it is.)

Like Sloman, these 1990s researchers asked how the different pathways can be integrated in various circumstances. But even they didn't really get the point of his work—which was that there are many visual pathways, not just two, and that these deal with many different types of information (e.g. transient versus long-lasting). Of course, Sloman was talking about neural computation, not neuroanatomy: there may or may not be distinct neuroanatomical pathways related to distinct types of function. (In low-level vision, it appears that there are.) The significant point here, however, is that his emphasis on the functional diversity of visual perception is still unusual. Hence my remark, above, that it has had only "something" of a revival.

Besides bearing comparison, up to a point, with Goodale and Milner's work, Sloman's current approach to vision fits in with recent psychological and neuro-scientific research on the internal emulation and anticipation of motor control, and on the perceptual feedback provided by motor action. For instance, on his website he describes his own work as being similar in spirit to the "emulation theory of representation" developed by the philosopher Grush (2004).

Grush's theory lies within the general tradition initiated 80 years ago by Kenneth Craik, wherein behaviour is guided by anticipatory mental/cerebral "models" of various kinds (Craik 1943; Boden 2006, p. 4.iv). Insofar as POPEYE was focussed on the use of a variety of mental representations, one could say that it, too, was situated in this tradition. But Sloman's focus has shifted. Today, he thinks of vision less as the analysis and interpretation of *representational structures* than as the analysis and interpretation of *informational processes*, with many different knowledge bases, and varying types of representation, acting (cooperating and competing) concurrently.

An even greater change has occurred in the attitude of AI researchers to work on emotion. The AI community has now woken up to the importance of emotion—although their interest is largely motivated by their wish to develop potentially lucrative gizmos. Current research on "companion robots" and the like tries to give AI systems recognition of, responsiveness to, and sometimes even simulation of, human emotions (Dautenhahn 2002; Wilks 2010).

Much of this work, it must be said—and has been said (Sloman 1999, 2001; cf. Picard 1999)—is shallow, for two reasons. First, many researchers still don't appreciate the degree of mental diversity that Sloman sketched years ago. Second, their primary aim (often) is not to understand how emotion functions in the control of other mental processes (including perception, thinking and motivation as well as action). Rather, it is to reassure, or even deceive, the human users of computer companions by providing a superficial appearance of emotional understanding and response on the part of the machine.

Whether gizmo-driven or not, however, the new interest in such matters has helped to draw international attention to Sloman's research on emotion, and on mental architecture in general. For instance, in 2002 DARPA invited him to take part in a small workshop on their new cognitive systems initiative, where his work was discussed in one of the introductory papers. Two years later, the AAAI held a cross-disciplinary symposium on "Architectures for Modelling Emotion". The EU also took an interest: it funded the 4-year CoSy project, begun in 2004 (Christensen et al. 2010), and the CogX project (2008–2012), led by Sloman's colleague Jeremy Wyatt (http://cogx.eu). These projects have produced a number of working models (research demos, not usable gizmos), but—like MINDER—these reflect only a small subset of Sloman's theoretical ideas.

Besides advising on large-scale (collaborative) projects such as these, and receiving many other invitations to speak, Sloman has been appointed as one of the leaders of a project addressing a "Grand Challenge" of British computing (see below). In this project, the architectural functions of emotion (in robots, as well as humans) are more important than the presentation of apparently emotional machine-companions to naïve users.

Even the philosophers—well, some of them—have woken up to the importance of Sloman's work. That's due in large part to a change in the philosophical background. (The *Zeitgeist,* again.) Analytically minded philosophers are now more ready to take account of scientific concepts and findings than they were in the mid-twentieth century. And some of them have specifically concerned themselves with AI (and sometimes with the concepts of computation and/or information), whether to defend or to reject its potential for illuminating the philosophy of mind. (The defenders include Grush, Jerry Fodor, Daniel Dennett, Paul Churchland, Steven Harnad, Andy Clark, Michael Wheeler, Brian Smith, Ronald Chrisley, Jack Copeland, Luciano Floridi, John Pollock, and myself; the sceptics include John Searle, Hubert Dreyfus, John Haugeland, Timothy van Gelder, Roger Penrose, Selmer Bringsjord and Ned Block.)

However, most philosophers are still largely ignorant of the AI details, so cannot engage with Sloman's work in a truly productive fashion. Moreover, the Turing Test, not to mention the Chinese Room (Searle 1980), still rears its ugly head far too often. The attempts of Sloman (1996a; 2002), and others, to defuse this ever-ticking bomb have not been taken to heart by his philosopher colleagues.

In addition, philosophers' ignorance is still often bolstered by philosophical principle. Dismissive charges of scientism are mounted by thinkers on the phenomenological side of the Anglo-Saxon/Continental, or realist/constructivist, divide (Boden 2006, 16.vi–viii). Unfortunately, these people—who don't even bother to read AI-based work–now comprise a larger fraction of the philosophical community than they did when Sloman was a young man.

The ideas of the later Wittgenstein (1953), who denied any place for computational (i.e. subpersonal) theories in psychology, have been used to attack cognitive science in general (e.g. Bennett and Hacker 2003). The Wittgensteinian philosopher Richard Rorty explicitly hoped for "the disappearance of psychology as a discipline distinct from neurology", including the demise of *computational* psychology (1979, p. 121). To make matters worse, several prominent writers originally trained in the analytic tradition, such as McDowell (1994), have adopted a phenomenological view according to which no naturalistic explanation of psychology is in principle possible. Even the founder of Turing-machine functionalism, who once urged philosophical comparisons between minds and computers, has reneged and turned to broadly constructivist accounts (Putnam 1967, 1982, 1988, 1997, 1999). In short, many philosophers today are just as loath to take Sloman's work seriously as they were in the 1970s.

Happily, more appreciation has come from other areas of cognitive science, as we've seen. If too many philosophers still steer clear of Sloman's work, because of their ignorance of AI and/or their suspicion of scientism, today's AI researchers do not.

In some AI-watchers' minds, to be sure, the appreciation pendulum has swung much too far in Sloman's favour—as he's the first to admit. Having become well known in AI circles for his work on emotions, he received an unexpected, and ridiculous, request. In his words: "I've even had someone from a US government-funded research centre in California phone me a couple of months ago [i.e. mid-2002] about the possibility of modelling emotional processes in terrorists. I told him it was beyond the state of the art. He told me I was the first person to say that: everyone else he contacted claimed to know how to do it (presumably hoping to attract research contracts)" (Sloman p.c.). (Probably, those other people weren't merely being opportunist, making promises they couldn't keep in order to board the Pentagon/Whitehall/EU band-waggons now funding research on "emotional" robots and "social" human–computer interactions. In addition, they didn't realize the depth and complexity of the mental-computational architecture that's required to generate emotional phenomena.)

Some people might accuse me, too, of valuing his work too highly. For in the final chapter of my recent book on the history of cognitive science, I listed a couple of dozen instances of research in this interdisciplinary field that I regard as especially promising—and said that if I were forced to choose only one, it would be Sloman's approach to integrated mental architecture (Boden 2006: p. 1449). Indeed, I'd already done that, when (on the 50th anniversary of the 1953 discovery of the double helix) the British Association for the Advancement of Science invited several people to write 200 words for their magazine *Science and Public Affairs* on "what discovery/advance/development in their field they think we'll be celebrating in 50 years' time". Others might well have prioritized a different item—or perhaps something not included on my list at all.

My choice, admittedly, was influenced by my own long-time interest, since high-school days, in personality and psychopathology (Boden 2006, Preface.ii). But it wasn't idiosyncratic. Two years later, the UK's computing community (of which AI researchers are only a subset) voted for "The Architecture of Brain and Mind" as one of the seven "Grand Challenges" for the future. (see http://www.uk.crc.org. uk/Grand_Challenges/index.cfm and http://www.cs.sir.ac.uk/gc5.) What's more, Sloman was appointed as a member of the five-man committee carrying this project forward.

In brief, many AI scientists, if not the committed gizmo-seekers, would endorse my valuation. They might do so while 'twinning' Sloman with Minsky, whose broadly similar research has considered architectural issues in more detail than is usual within AI (Minsky 1985, 2007). They might even judge Minsky's work above Sloman's, especially if they have little interest in philosophical questions. But I'd be very surprised if anyone seriously concerned with the nature of minds as a whole were not to appreciate Sloman's contribution.

2.7 Coda

I've focussed only on Sloman's own intellectual work. But I must also mention his importance as an instigator of AI research and education, in the UK and elsewhere.

Having spent a year at the University of Edinburgh's Machine Intelligence Unit, with Donald Michie and Bernard Meltzer (and many younger researchers in AI), he was a main driver in setting up the Cognitive Studies Programme at the University of Sussex in the early-1970s. The other founders of this interdisciplinary venture included the charismatic Max Clowes (an imaginative early researcher in computer vision: Clowes 1967, 1969, 1971), Alistair Chalmers (a highly computer-literate social psychologist), and myself (already using AI to understand the mind: Boden 1965, 1970, 1972, 1973). As the world's first academic programme to integrate AI with philosophy, psychology and linguistics, COGS became internationally recognized, and widely influential.

As part of this educational project, Sloman (with colleagues such as John Gibson and Steven Hardy) developed the highly user-friendly POPLOG programming system, soon to be used by other universities and by various commercial institutions. Later, he took a key role in the UK's government-backed Alvey Programme, which encouraged AI knowledge transfer between academia and industry (Boden 2006, p. 11.iv-v). Partly due to his influence, the Alvey remit was broadened from logic programming and expert systems to include vision and neural computation too.

Over the years, a number of other important advisory/administrative roles in the UK's and Europe's computing community followed. For instance, in 2003 he was one of six AI researchers consulted about the new EU Cognitive Systems Initiative. And since 2009 he has been heavily involved in the UK's Computing at School initiative—trying, among other things, to make this focussed more on using AI/computing to understand the mind and less on using or generating gizmos, a.k.a. apps (http://www.computingatschool.org.uk).

In short, even setting aside his own research, Sloman has been—and continues to be—prominent in virtue of the insightful advice he has given on AI and computing in the UK and beyond.

Finally, Sloman as a person. He could not have achieved the degree of intellectual leadership he has exercised at many different levels without being someone who drew affection, as well as respect, from others. That affection was largely earned by his own unfailing respectfulness for those who came in contact with him. Add to this, his exceptional generosity in helping his colleagues and students—a generosity that devoured precious time that could have been spent more selfishly.

I myself have benefitted from this on various occasions. My book *Artificial Intelligence and Natural Man* (1977) was much improved by his advice, and contained this seemingly bland but actually heartfelt Acknowledgment: "I am deeply grateful to Aaron Sloman for his careful reading of the draft manuscript, and for many conversations on related topics". In addition, he has helped me countless times, with admirable patience, to cope with the technology—as he has done for many others, too. In the 50 years that I've known him (since 1962), Aaron has been my most intellectually stimulating colleague, and a very dear friend.

References

Anderson JR (1983) The architecture of cognition. Harvard University Press, Cambridge

Baker S (2011) Final Jeopardy: man vs. machine and the quest to know everything. Houghton Mifflin Harcourt, Boston

Beaudoin LP (1994) Goal processing in autonomous agents. Ph.D. thesis, School of Computer Science, University of Birmingham. Available at http://www.cs.bham.ac.uk/research/cogaff/

Bennett MR, Hacker PMS (2003) Philosophical foundations of neuroscience. Blackwell, Oxford

Boden MA (1959) In reply to hart and hampshire. Mind (NS) 68:256–260

Boden MA (1965) McDougall revisited. J Pers 33:1–19. [Reprinted in Boden MA (1981) Minds and mechanisms: philosophical psychology and computational models. Cornell University Press, Ithaca, pp. 192–208]

Boden MA (1969) Machine perception. Philos Q 19:32–45

Boden MA (1970) Intentionality and physical systems. Philos Sci 37:200–214

Boden MA (1972) Purposive explanation in psychology. Harvard University Press, Cambridge

Boden MA (1973) How artificial is artificial intelligence? Br J Philos Sci 24:61–72

Boden MA (1977) Artificial intelligence and natural man. Basic Books, New York. (2nd edn., expanded, 1987. MIT Press, London; Basic Books, New York)

Boden MA (ed) (1990) The philosophy of artificial intelligence. Oxford University Press, Oxford

Boden MA (2006) Mind as machine: a history of cognitive science. Clarendon/Oxford University Press, Oxford

Brachman RJ, Levesque HJ (eds) (1985) Readings in knowledge representation. Morgan Kauffman, Los Altos

Brooks RA (1990) Elephants don't play chess. Robot Auton Syst 6:3–15

Christensen HI, Kruijff G-JM, Wyatt JL (eds.) (2010) Cognitive systems. Cognitive systems monographs vol 8. Springer, Berlin

Clark AJ (2013) Whatever next? predictive brains, situated agents, and the future of cognitive science. Behav Brain Sci (in press)

Clowes MB (1967) Perception, picture processing, and computers. In: Collins NL, Michie DM (eds) Machine intelligence 1. Edinburgh University Press, Edinburgh, pp 181–197

Clowes MB (1969) Pictorial relationships-a syntactic approach. In: Meltzer B, Michie DM (eds) Machine intelligence 4. Edinburgh University Press, Edinburgh, pp 361–383

Clowes MB (1971) On seeing things. Artif Intell 2:79–116

Craik KJW (1943) The nature of explanation. Cambridge University Press, Cambridge

Crevier D (1993) Ai: the tumultuous history of the search for artificial intelligence. Basic Books, New York

Damasio AR (1994) Descartes' error: emotion, reason, and the human brain. Putnam, New York

Dautenhahn K (ed) (2002) Socially intelligent agents: creating relationships with computers and robots. Kluwer Academic, Boston

Dennett DC (1988) Quining qualia. In: Marcel A, Bisiach E (eds) Consciousness in contemporary science. Oxford University Press, Oxford, pp 42–77

Dennett DC (1991) Consciousness explained. Allen Lane, London

Fisher M (1990) Personal love. Duckworth, London

Frege G (1884/1950) The foundations of arithmetic (trans. Austin JL). Oxford University Press, Oxford

Frith U (1989/2003) Autism: explaining the Enigma, 2nd edn. Blackwell, Oxford (rev. 2003)

Gandy R (1996) Human versus mechanical intelligence. In: Millican PJR, Clark AJ (eds) Machines and thought: the legacy of alan turing, vol I. Oxford University Press, Oxford, pp 125–136

Gibson JJ (1966) THe senses considered as perceptual systems. Greenwood Press, Westport

Goodale MA, Humphrey GK (1998) The objects of action and perception. Cognition 67:181–207

Goodale MA, Milner AD (1992) Separate visual pathways for perception and action. Trends Neurosci 13:20–23

Goodale MA, Milner AD (2003) Sight unseen: an exploration of conscious and unconscious vision. Oxford University Press, Oxford

Grush R (2004) The emulation theory of representation. Behav Brain Sci 27:377–442

Laird JE, Newell A, Rosenbloom P (1987) Soar: an architecture for general intelligence. Artif Intell 33:1–64

McCarthy J, Hayes PJ (1969) Some philosophical problems from the standpoint of artificial intelligence. In: Meltzer B, Michie DM (eds) Machine intelligence 4. Edinburgh University Press, Edinburgh, pp 463–502

McDowell J (1994) Mind and world. Harvard University Press, Cambridge

Marr DC (1974a) The computation of lightness by the primate retina. Vision 14:1377–1388

Marr DC (1974b) A note on the computation of binocular disparity in a symbolic, low-level visual processor. MIT AI-Lab Memo no. 327, Cambridge. [Reprinted in Vaina L (ed) From the retina to the neocortex: selected papers of David Marr. Birkhauser, Boston, pp 231–238 (1991)

Marr DC (1975a) Analyzing natural images: a computational theory of texture vision.AI Memo 334. MIT AI Lab., Cambridge

Marr DC (1975b) Early processing of visual information. AI Memo 340. MIT AI Lab, Cambridge. [December 1975. Officially published in Philos Trans Roy Soc B 275:483–524 (1976)

Marr DC (1975c) Approaches to biological information processing. Science 190:875–876

Marr DC, Hildreth E (1980) Theory of edge-detection. Proc Roy Soc B 207:187–217

Marr DC, Nishihara HK (1978) Visual information processing: artificial intelligence and the sensorium of sight. Technol Rev 81:2–23

Marr DC, Poggio T (1976) Cooperative computation of stereo disparity. Science 194:283–287

Marr DC, Poggio T (1977) From understanding computation to understanding neural circuitry. Neurosci Res Prog Bull 15:470–488

Marr DC, Poggio T (1979) A computational theory of human stereo vision. Proc Roy Soc B 204:301–328

Milner AD, Goodale MA (1993) Visual pathways to perception and action. In: Hicks TP, Molotchnikoff S, Ono T (eds) Progress in brain research, vol 95. Elsevier, Amsterdam, pp 317–337

Minsky ML (1965) Matter, mind, and models. In: Proceedings of the international federation of information processing congress, vol 1. Spartan, Washington, pp 45–49

Minsky ML (1985) The society of mind. Simon & Schuster, New York

Minsky ML (2007) The emotion machine: commonsense thinking, artificial intelligence, and the future of human mind.

Newell A (1973) Artificial intelligence and the concept of mind. In: Schank RC, Colby KM (eds) Computer models of thought and language. Freeman, San Francisco, pp 1–60

Newell A, Simon HA (1961) GPS—A program that simulates human thought. In: Billing H (ed) Lernende Automaten. Oldenbourg, Munich, pp 109–124. [Reprinted in Feigenbaum EA, Feldman JA (eds) (1963) Computers and thought. McGraw-Hill, pp 279–293]

Picard RW (1997) Affective computing. MIT Press, Cambridge

Picard RW (1999) Response to sloman's review of affective computing. AI Magazine, 20/1 (March), pp 134–137

Putnam H (1967) The nature of mental states. First published as 'Psychological Predicates' in Capitan WH, Merrill D (eds) Art, mind, and religion. University of Pittsburgh Press, Pittsburgh, pp 37–48. [Reprinted in Putnam H (1975) Mind, language, and reality: philosophical papers, vol 2. Cambridge University Press, Cambridge, pp 429–440]

Putnam H (1982) Why there isn't a ready-made world. Synthese 51:141–167

Putnam H (1988) Representation and reality. MIT Press, Cambridge

Putnam H (1997) Functionalism: cognitive science or science fiction? In: Johnson DM, Erneling CE (eds) the future of the cognitive revolution. Oxford University Press, Oxford, pp 32–44

Putnam H (1999) The threefold cord: mind, body, and world. Columbia University Press, New York

Rorty R (1979) Philosophy and the mirror of nature. Princeton University Press, Princeton

Ryle G (1949) The concept of mind. Hutchinson's University Library, London

Searle JR (1980) Minds, brains, and programs. Behavior Brain Sci 3(3):417–457

Simon HA (1962) The architecture of complexity. Proc Am Philos Soc 106(1962):467–482

Simon HA (1967) Motivational and emotional controls of cognition. Psychol Rev 74:29–39

Simon HA (1969) The sciences of the artificial. The Karl Taylor compton lectures. MIT Press, Cambridge. (2nd and 3rd edns. 1981 and 1996)

Sloman A (n.d.) The CogAff group's website: http://www.cs.bham.ac.uk/research/cogaff

Sloman A (1971) Interactions between philosophy and artificial intelligence: the role of intuition and non-logical reasoning in intelligence. Artif Intell 2:209–225

Sloman A (1974) Physicalism and the Bogey of determinism. In: Brown SC (ed) Philosophy of psychology. Macmillan, London, pp 283–304

Sloman A (1975) Afterthoughts on analogical representation. In: Schank RC, Nash-Webber BL (eds) (1985) Theoretical issues in natural language processing: an interdisciplinary workshop in computational linguistics, psychology, linguistics, and artificial intelligence, Cambridge, Mass., 10–13 June. (Arlington, Va.: Association for computational linguistics), pp 164–168. Reprinted in Brachman and Levesque, pp 431–39

Sloman A (1978) The computer revolution in philosophy: philosophy, science, and models of mind. Harvester Press, Brighton. Out of print but available—and continually updated—online at http://www.cs.bham.ac.uk/research/cogaff/crp/

Sloman A (1982) Towards a grammar of emotions. New Univ Q 36:230–238

Sloman A (1983) Image interpretation: the way ahead? In: Braddick OJ, Sleigh AC (eds) Physical and biological processing of images. Springer-Verlag, New York, pp 380–40

Sloman A (1986) What sorts of machine can understand the symbols they use?. In: Proceedings of the Aristotelian society, Supplementary 60:61–80

Sloman A (1987a) Motives, mechanisms, and emotions. Cogn Emot 1: 217–233. (Reprinted in Boden 1990:231–247)

Sloman A (1987b) Reference without causal links. In: du Boulay JBH, Hogg D, Steels L (eds) Advances in artificial intelligence—II. North Holland, Dordrecht, pp 369–381

Sloman A (1989) On designing a visual system: towards a gibsonian computational model of vision. J Exp Theor AI 1:289–337

Sloman A (1992) The emperor's real mind review of roger penrose's the emperor's new mind: concerning computers minds and the laws of physics. Artif Intell 56:355–396

Sloman A (1993) The Mind as a Control System. In: Hookway C, Peterson D (eds) Philosophy and the Cognitive Sciences. Cambridge University Press, Cambridge, pp 69–110

Sloman A (1995) Sim_Agent help-file. Available at ftp://ftp.cs.bham.ac.uk/pub/dist/poplog/sim/help/sim_agent. See also 'Sim_agent web-page, Available at http://www.cs.bhma.ac.uk/axs/cog_affect/sim_agent.html

Sloman A (1996a) Beyond turing equivalence. In: Millican PJR, Clark AJ (eds) Machines and thought: the legacy of alan turing, vol 1. Oxford University Press, Oxford, pp 179–220

Sloman A (1996b) Towards a general theory of representations. In: Peterson DM (ed) Forms of representation: an interdisciplinary theme for cognitive science. Intellect Books, Exeter, pp 118–140

Sloman A (1996c) Actual possibilities. In: Aiello LC, Shapiro SC (eds) Principles of knowledge representation and reasoning. In: Proceedings of the fifth international conference (KR '96). Morgan Kaufmann, San Francisco, pp 627–638

Sloman A (1998) Ekman, damasio, descartes, alarms and meta-management. In: Proceedings of the international conference on systems, man, and cybernetics (SMC98), IEEE Press, San Diego, pp 2652–2657

Sloman A (1999) Review of [R. Picard's] affective computing. AI Magazine 20:1 (March), pp 127–133

Sloman A (2000) Architectural requirements for human-like agents both natural and artificial. (What sorts of machines can love?). In: Dautenhahn K (ed) Human cognition and social agent technology: advances in consciousness research. John Benjamins, Amsterdam, pp 163–195

Sloman A (2001) Beyond shallow models of emotion. Cogn Process Int Q Cogn Sci 2:177–198

Sloman A (2002) The irrelevance of turing machines to artificial intelligence. In: Scheutz M (ed) Computationalism: new directions. MIT Press, Cambridge, pp 87–127

Sloman A (2003) How many separately evolved emotional beasties live within us?. In: Trappl R, Petta P, Payr S (eds) Emotions in humans and artifacts. MIT Press, Cambridge, pp 29–96

Sloman A (2009a) Requirements for a fully deliberative architecture (or component of an architecture). Available on the CogAff website: http://www.cs.bham.ac.uk/research/projects/cog-aff

Sloman A (2009b) Some requirements for human-like robots: why the recent over-emphasis on embodiment has held up progress. In: Sendhoff B, Koerner E, Sporns O, Ritter H, Doya K (eds) Creating brain-like intelligence: from basic principles to complex intelligent systems. Lecture notes in computer science, vol 5436. Springer-Verlag, Berlin, pp 248–277

Sloman A (2013) What else can brains do?: commentary on A. Clark's whatever next? Behav Brain Sci (in press)

Sloman A, Chrisley RL (2003) Virtual machines and consciousness. In: Holland O (ed) Machine consciousness. Imprint Academic, Exeter, pp 133–172

Sloman A, Croucher M (1981) Why robots will have emotions. In: Proceedings of the Seventh international joint conference on artificial intelligence (Vancouver), pp 197–202

Sloman A, Poli R (1995) Sim_Agent: A toolkit for exloring agent designs. In: Wooldridge M, Muller J-P, Tambe M (eds) Intelligent Agents, vol ii. Springer-Verlag, Berlin, pp 392–407

Squires R (1970) Are dispositions lost causes? Analysis 31:15–18

Turing AM (1950) Computing machinery and intelligence. Mind 59: 433–460. [(Reprinted in (Boden 1990: 40–66)]

Wilks YA (ed) (2010) Close engagements with artificial companions: key social, psychological ethical and design issues. John Benjamins, Amsterdam

Wittgenstein L (1953) Philosophical investigations (Trans. Anscombe GEM). Blackwell, Oxford

Wright IP (1997) Emotional agents. Ph.D. thesis, School of Computer Science, University of Birmingham. Available at http://www.cs.bham.ac.uk/research/cogaff/

Wright IP, Sloman A (1997) MINDER1: An implementation of a protoemotional agent architecture. Technical Report CSRP-97-1, University of Birmingham, School of Computer Science. Available at ftp://ftp.cs.bham.ac.uk/pub/tech-reports/1997/CSRP-97-01.ps.gz

Wright I, Sloman A, Beaudoin L (1996) Towards a design-based analysis of emotional episodes. Philos Psychiatry Psychol 3(2):101–126

Chapter 3
Losing Control Within the H-Cogaff Architecture

Dean Petters

3.1 Introduction

This papers aims to use an analysis of multiple but related psychological phenomena to work towards developing deeper models of emotion and richer cognitive models in general. These phenomena have in common some lapse or loss of control. Many existing ways to model emotion rely upon shallow models which in turn rely upon easily measurable or reportable behaviours and other phenomena in various test situations, such as brain images or physiological data. As Sloman (2001a) notes: "*A desirable but rarely achieved type of depth in an explanatory theory is having a model which accounts for a wide range of phenomena. One of the reasons for shallowness in psychological theories is consideration of too small a variety of cases.*" The examples where control is lost which are considered in this paper are a diverse collection that vary in duration from momentary episodes of fear to the development of emotional attachments and experiences of grief which can be greatly extended in time. They vary in intensity from how a school pupil is motivated to greater success in school by a wish for their parents to gain pride in them to examples of murderous and uncontrolled rage. They also vary from phenomena which are described by a central mechanism, such as self-deception resulting from Freudian Repression, to the multiple distributed mechanisms which von Hippel and Trivers (2011) invoke in the formation of self-deception. These phenomena are integrated by explaining them all in terms of a single information processing architecture. In this integration deeper theoretical architecture-based concepts are introduced such as interrupts or disturbances to processing and changes in the locus of control or access to information are introduced. These concepts are then used to identify subsets of

D. Petters (✉)
School of Social Sciences, University of Northampton, Northampton, UK
e-mail: dean.petters@northampton.ac.uk

J. L. Wyatt et al. (eds.), *From Animals to Robots and Back: Reflections on Hard Problems in the Study of Cognition*, Cognitive Systems Monographs 22, DOI: 10.1007/978-3-319-06614-1_3, © Springer International Publishing Switzerland 2014

the phenomena related to loss of control. Some phenomena may be best described by central processes within an architecture whereas others might focus on relations with the perceived physical or social environment (Sloman 2001a).

3.2 What Are Emotions?

In 'The Expression of Emotions' (Darwin1872), Darwin presented relatively shallow behavioural criteria for emotions based upon the production of facial expressions such as blushing or frowning. This approach to describing typical emotional behaviours has been updated in the work of Ekman (2003). Other relatively shallow measurable or reportable criteria for emotions include: physiological measures such as increased heart rate; the activity of specific regions of the brain; the introspectable and reportable experience of bodily changes or desires, such as wanting to run away, or to hurt someone; the experience of interpreting and labelling of situations which trigger emotions, such as appraising a loud noise as a threat; and typical behavioural responses to emotions such as fighting or running away (Sloman 2001a). These kinds of shallow criteria for emotion can be organized into four categories that describe an emotion as: (1) a neurally implemented response, (2) with a conscious feeling that possesses sensory qualities, (3) including a cognitive process of interpretation and (4) resulting in a behavioural response. In his recent book 'What is Emotion?' Jerome Kagan states that the belief that human emotion is constituted of these four categories is a widely held view amongst psychology researchers. However, Kagan notes that states may exist which possess some but not all of these criteria and: *"scholars vary [...] in the significance they award to each of these four components"* (Kagan 2007, p. 23).

One of the problems with using shallow criteria for emotions based on emotional components is that they underspecify emotional phenomena. Kagan himself notes that the behavioural component is not a strong criterion for confirming an emotion has occurred. As he states, *"an emotion need not be accompanied by any behaviour; that is why the concept of 'emotion regulation' was invented"* (Kagan 2007, p. 192). Colombetti and Thompson (2008) put forward a critique of emotion research based on how the cognitive component of emotion is currently understood. They suggest that much research in emotion is based upon processes of cognitive appraisal that treat cognitive and bodily events as separate processes. They argue that this is because the cognitive appraisal component of emotion has been conceptualised as an *"abstract, intellectual, 'heady' process separate from bodily events"* (Colombetti and Thompson 2008, p. 45). So according to this embodied view emotional experience may involve an interpretation of events but this interpretation need not involve processes which are mediated by discrete representations which are acted upon by rule manipulating systems. If Colombetti and Thompson (2008) are correct then much more representationally diverse mechanisms for emotion evaluation and appraisal may need to be considered.

The emotional experience component is also not a clear criterion for validating emotions. Not all agree with Kagan's (2007) view that emotions are by definition conscious. For example, people can be in emotional states like jealousy of which they are not conscious, or in long-term emotional states of which they are only intermittently conscious like grief, which continue to predispose individuals to take particular actions even at moments they are not conscious of this emotion. As Sloman (2001a) also notes, people may be unaware of their enjoyment when engrossed in a game of football or watching an opera with attention fully engaged. The requirement for a neural implementation for emotion is challenged by researchers who take a functional view of emotions (Arbib and Fellous 2004). For example, Evans argues that if we say that emotions have to be mediated by neural structures similar to those found in humans then neither computers nor intelligent aliens with exotic brains could ever to be said to have emotions and that *"this is a very parochial view of emotion"* (Evans 2003, p. 102).

What the above discussion of emotional components has shown is that emotions can exist in the absence of one or more of these components. Instead of defining emotions in these terms we can form an integrative framework that considers for particular emotional states which of the four components detailed above are relevant and how those that are relevant are instantiated. In the remainder of the paper, Sect. 3.3.1 reviews possible frameworks which might support an integration of the various psychological phenomena related to loss of control being considered in this paper. Section 3.3.2 showing how the Cogaff schema and HCogAff architecture can incorporate a range of phenomena associated with different kinds of emotion including loss of control. Finally, Sect. 3.4 provides a description of the six examples of loss of control which are the focus of this paper.

3.3 From Shallow to Deep Models of Emotion

3.3.1 Emotions Theories Organised by Components and Phases

This section presents an analysis of diverse emotional episodes which have in common that some element of self-control is lost. An integrated explanatory framework for multiple phenomena should be able to include diverse information structures, mechanisms and processes. Such a broad and rich conceptual toolkit can then support deep models of emotional phenomena. One way to bring together a diversity of approaches is to bring together and combine multiple competing theories of emotion. By aggregating together multiple emotion theories we might hope to gain the breadth and richness required for deep models of self-control and the loss of self-control. Scherer (2010, pp. 10–15) is concerned with assessing the differential utility of emotion theories for dynamic modelling of emotional phenomena. In pursuit of this, he provides a review of existing emotion theories by grouping a large number of extant theories into eight major 'families' of emotion theory. In Scherer's approach he char-

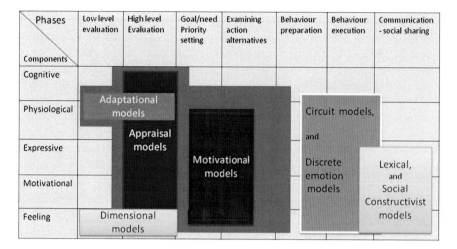

Phases Components	Low level evaluation	High level Evaluation	Goal/need Priority setting	Examining action alternatives	Behaviour preparation	Behaviour execution	Communication - social sharing
Cognitive							
Physiological	Adaptational models				Circuit models,		
		Appraisal models			and		
Expressive			Motivational models		Discrete emotion models	Lexical, and Social Constructivist models	
Motivational							
Feeling	Dimensional models						

Fig. 3.1 An adapted version of Scherer's mapping of competing emotion theories onto a grid defined by five emotion components and seven emotional phases. (adapted from Scherer (2010)). This diagram shows that the eight competing emotion theories can be organised into three larger theory-clusters described by Scherer (2010, pp. 14–15) so that: adaptational, appraisal and motivation models form an 'evaluation and decision' set of theories with their focus on explaining early phases of emotional episodes (this cluster surrounded by a *dark grey background*); circuit and discrete emotion models together form a 'preparation and execution' set of theories focused on later phases in emotional episodes; and dimensional, lexical, and social constructivist models, form a 'subjective and social' set of theories which are characterised by having less focus on internal 'decision' processes and more on internal subjective feelings and the external social environment

acterises emotion theories according to two set of criteria. The first set of criteria are categorising emotion theories according to emotional components. The same four components for emotion are used as described above by Kagan (neural implementation, conscious feelings, cognitive processes, and behavioural responses) with the addition of motivation as a fifth emotional component. The second set of criteria are the seven phases of an emotional episode, which according to Scherer are: low-level evaluation; high-level evaluation; goal/need priority setting; examining action alternatives; behaviour preparation; behaviour execution; and communication and social sharing. Figure 3.1 shows how eight classes of emotion theory can be mapped onto the 35 subdivisions of a grid formed by having 5 emotion components as one axis and 7 emotion phases the other axis. The eight 'families' of emotion theory categorised according to these criteria are:

- Adaptational theories view emotion as possessing an important adaptive function and are centred around both low and high-level evaluation phases (particularly on fear inducing stimuli) and focus on describing and explaining cognitive and physiological components of emotion. LeDoux's (1996) model of fear is an example of this type of theory.

- Dimensional theories differentiate emotions on their position on two dimensions: pleasantness–unpleasantness and arousal. They focus on the feeling emotional component.
- Appraisal theories focus on the cognitive component of emotion whilst also being linked to physiological, expressive and motivational components. They are centred around the high-level evaluation phase of emotions.
- Motivational theories argue that emotions are derived from evolutionary motivational primitives and whilst focusing across physiological, expressive and motivational components are centred on the goal/need priority setting and examining action alternatives phases of emotion.
- Circuit theories focus on expressive and motivational components of emotion and are particularly influenced by pathways and circuits in the brain that are described at a neural level. They explain events in the behaviour preparation and behaviour execution emotional phases.
- Discrete (basic) theories of emotion include ideas from Darwin (1872) and Ekman (2003). They posit that there exist a small number of discrete basic emotions and are focused on physiological, expressive, motivational and feeling components of emotion and are centred around the behaviour preparation and behaviour execution emotional phases.
- Lexical theories assume that the semantic structure of how emotion terms are used will reveal the underlying organisation and determinants of emotions. These theories are focused at the expressive, motivational and feeling components of emotion and are centred around the behaviour execution and communication and social sharing emotional phases.
- Social Constructivist theories are favoured by sociologists and anthropologists who hold that meaning in emotion is socioculturally determined. As with lexical theories, these theories are focused at the expressive, motivational and feeling components of emotion and are centred around the behaviour execution and communication and social sharing emotional phases.

After categorising emotion theories into eight major 'families' Scherer (2010, pp. 14–15) continues his ambitious classification system by undertaking a further clustering by looking at overlap between these eight families. So Fig. 3.1 shows that adaptational, appraisal and motivation models form an 'evaluation and decision' cluster of theories that focus on the initial evaluation, judgement and decision-making phases of an emotional episode; circuit and discrete emotion models together form a 'preparation and execution' cluster whose focus starts where the first cluster finishes. The third cluster brings together dimensional, lexical, and social constructivist models. These theories do not have a shared focus based upon a particular phase as dimensional theories focus on early stages of emotional episodes and lexical and social constructivist theories on later written and interpersonal interactions. However, these 'subjective and social' models do focus more on internal subjective feelings and the external social environment.

Scherer suggests that when choosing a theory as a basis for computational modelling of emotion, we can integrate the theories serially—by linking up theories in

	LONGER TERM	INTERMEDIATE	SHORTER TERM
Fig. 3.2 Classes of semantic control state, which are compared with respect to the approximate duration that each class of control state may exist as a disposition within an architecture (Adapted from Sloman (1995) and Petters (2006))	Personality, Temperament, Attitudes, Skills, Emotions such as love, grief, Attachment style	Moods, Beliefs, Preferences, Emotions such as joy, fear, Intentions, Plans, Desires	neural and physical events,

early to late phases of an emotional episode. What this 'chaining' of theories does not give is an overarching theory of all the non-emotional information processing that fills in the gaps in a complete information processing architecture. In assessing Scherer's classification system as a framework for supporting deep explanations of self-control and the loss of control a key question is: does this system provide the kind of 'wrap-around' conceptualisation necessary? What this means is that self-control may not be located in any single component or phase but occurs between and around components and phases. So initial perceptions can affect later behaviour and the converse can occur. For example, behavioural outcomes can trigger self-control of how situations are perceived or judged. An explanatory framework used to support deep models of self-control and the loss of such control therefore needs to include connections and interrelationships between processes occurring across the 35 elements of Scherer's grid in Fig. 3.1. These may not occur in a serial 'pipeline' from early to late phases of an emotional episode. The wide-ranging connections between early and late processes and different emotional components can be the basis of emotion regulation and therefore the basis of self-control. Figure 3.1 highlights that Scherer's three 'super-family' clusters of emotion theory have a gap between 'evaluation and decision' focussed theories and 'preparation and execution' focussed theories. There is not a single 'super-family' of theories which captures interrelationships of self-reflection and control between all emotional phases and all emotional components in a complete and 'wrapped-around' integrated manner. What we therefore require is an approach which does not force us to decide on which phases or components to focus on at the outset but provide a natural coverage of all components, phases, and the possible control processes between these subdivisions. Such a framework would include very long-term dispositions such as attachment status as well as shorter term control states like momentary anger and fear (see Fig. 3.2). Aggregation of emotion theories does not on its own provide the integrative links we require to model self-control and the loss of control, perhaps because they understandably focus mainly on emotional phenomena. So non-emotional processes are therefore more peripheral and given less attention in each individual theory. So when aggregation occurs between emotional theories these non-emotional processes are not included and so not available' to provide a 'glue' to stick all the emotional pieces together. The next section reviews the Cognition and Affect approach which situates emotions within a broader information processing framework where non-emotional processes are considered alongside affective and emotional processes.

3.3.2 Emotions Within the CogAff Schema and HCogAff Architecture

Sloman (2001a) portrays the varying approaches in how emotion is researched as similar to the contradictory opinions expressed by the proverbial ten blind men each trying to say what an elephant is on the basis of feeling only a small part of it. Instead of arguing over which description is right, the ideal solution is for the men describing an elephant to attempt to describe the whole elephant. A way to bring about theoretical integration in emotion research and work towards describing 'the whole elephant' is to form computational models which are based upon cognitive architectures which possess a rich internal structure and show how different emotional classes and phenomena can be produced.

Sloman (2000, 2001a, 2002) has set out the CogAff schema (Fig. 3.3a) and the HCogAff architecture (Fig. 3.3b) as conceptual tools which facilitate integrating diverse information processing structures, and mechanisms, and explanations of behavioural phenomena. The CogAff schema is a systematic architectural framework which is intended only as a first approximation to summarising layers of control and cognition produced by evolution. Although there is as yet no agreed conceptual framework for describing architectures, the CogAff schema makes some high-level distinctions (Sloman 2001a). The CogAff schema (Fig. 3.3a) organises information processing by overlaying three layers (reactive; deliberative; and meta-management) and three columns (perception, central processing and action). The HCogAff architecture is a special case (or subclass) of CogAff which has not been implemented in its entirety though the production of some subsets have been accomplished (Petters 2006; Wright 1997). A very important distinction in the HCogAff architecture is between non-attentive reactive or perceptual processes and the attentive processes which occur within the variable attention filter in Fig. 3.3b. Motive generactivators operate in parallel in the non-attentive reactive component of HCogAff and act to generate to activate motives. They are triggered by internal and external events. They can be thought of as 'scouring' the world for their firing conditions (Wright et al. 1996). When these conditions are met a motivator is constructed which may 'surface' above the attentional filter and be operated upon by processes in the deliberative or meta-management levels. Amongst the processes generated by motivators are: evaluations, prioritisation, selection, expansion into plans, plan execution, plan suspension, prediction, and conflict detection. Management processes can form and then operate upon explicit representations of options before selecting options for further deliberation (Wright et al. 1996). Representations in the deliberative layer can be formed with compositional semantics that allow a first-order ontology to be formed which includes future possible actions. The meta level can operate using

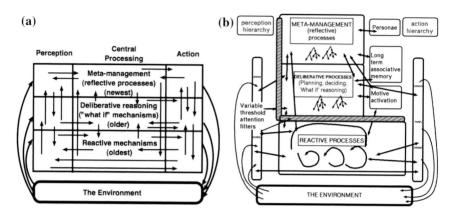

Fig. 3.3 The CogAff framework (thanks to Aaron Sloman for permission to use these graphics)

meta-semantics which allows a second-order ontology to be formed, which refers to the first- order ontology used within the deliberative level (Sloman and Chrisley 2005).

Although the HCogAff architecture has a large-scale structure which endures over time, there is constant relocating and transforming of motivators which is termed circulation. As Wright et al. (1996) notes, useful control states become more influential and 'percolate' up a hierarchy of dispositional control states. Ineffective motivators wither away in influence. One important process is 'diffusion', in which the impact of a major motivator leads it to become gradually distributed in myriad control states which can include new motive generators, plans, preferences, predictive models, reflexes and automatic responses (wright et al. 1996). Meta-management attempts to influence these numerous processes but some are more controllable than others. In summary, the HCogAff architecture is a control system with a rich collection of control states including a numerous ways in which some processes manage or influence other processes (wright et al. 1996).

3.3.2.1 Mapping Emotional Phases to the HCogAff Architecture

A requirement for a integrative theoretical framework for emotion with comprehensive coverage is that it can represent processes occurring in all the seven emotion phases set out by Scherer (2010). The HCogAff architecture (Fig. 3.3b) represents all phases of information processing resulting from sensation and perception, to central processing, and then motor control (Sloman 2000, 2002). It should therefore naturally cover all the seven emotional phases described by Scherer (2010). For example, *Low level evaluation processes* might be mapped to the HCogAff architecture's perceptual subsystem of the reactive layer and *High-level evaluation* can be mapped to the perceptual and central processing subsystems of the deliberative layer. *Goal/need priority setting* and *examining action alternatives* might be mapped to central processing and action subsystems of the reactive and deliberative layer. *Behaviour preparation, Behaviour execution*, and *Communication and social sharing* might all be mapped to different levels of HCogAff architecture's central processing and action subsystems. Further work is needed to provide more details of the possible mappings, perhaps integrating evidence from brain imaging and neuropsychology with computational modelling and behavioural studies.

3.3.2.2 Mapping Emotional Components to the HCogAff Architecture

An integrative theoretical framework should also be able to focus on the four emotional components described by Kagan (2007). The distinction between reactive and deliberative processes can help capture both the neural and cognitive emotional components within a single architecture. The action subsystems at the reactive, deliberative and meta processing levels provide a focus on the behavioural component of an emotion. Of the emotional components presented by Kagan (2007) and Scherer (2010) representing subjective emotional feelings provides a significant challenge for any information processing framework. However, Sloman and Chrisley (2003) argue that an implementation of the HCogAff architecture could develop an inherently private and incommunicable ontology for referring to its own perceptual contents and other internal states. That ontology would be produced by self-organising classification mechanisms and may explain aspects of qualia relevant to subjective emotional feelings (Sloman 2010).

3.4 Six Ways to Lose Control

3.4.1 Emotions and Real-Time Responses

When faced with a decision, it takes time to think through all the evidence and options in a completely rational manner. Sometimes an organism or artefact will not have enough time to make a fully rational decision before disaster is upon it and a less considered but faster decision may have been greatly preferable. The idea of emotions as interrupts to ongoing processing was developed by Simon (1967). The central argument in Simon (1967) is that human information processing must cope with multiple needs in an unpredictable environment. These requirements can be met by two classes of mechanism. First, there are goal terminating mechanisms which permit multiple goals to be processed serially without any one need monopolising the processor. So processing for a particular need is stopped when that need is achieved, when too much time has been spent attempting to achieve the need, or progress towards achieving the need is too slow. Simon shows that these mechanisms allow serial processors to respond to a multiplicity of motives existing within a control hierarchy at the same time without any requirement for special mechanisms that represent affect or emotion. However, these mechanisms are inadequate if the system's processing speed is fixed and there are real-time demands on the system. Then provision must be made for an interrupt system. There are two requirements for an interrupt system: processing towards the main goal must go on continuously in parallel with processing that enables the system to notice when interrupts are required; and when real-time needs of high priority are noticed the noticing program must be capable of setting aside ongoing processing and substituting with a quick response.

LeDoux (1996, p. 164) describes in detail the neural basis for fear interruptions operating in rat and human brains in a similar manner to how Simon's description is set out above. LeDoux characterises two different routes to appraisal of possibly threatening objects from the sensory thalamus to the amygdala. These are labelled as a 'low road' and a 'high road'. The low road is a 'quick and dirty' processing pathway that bypasses the cortex. This direct thalamo-amygdala route only provides a crude representation of the threatening object but is very fast. The high road is a slower route to action via the cortex. This route eventually ends up in the amygdala but on the way has allowed cortical processing to provide recognition of the object. The high road can be blocked and overridden in an 'emotional hijacking' when the low road takes control—so that we can start responding to dangerous stimuli before we know what they are. As LeDoux notes, this is very useful in dangerous situations but can result in emotional responses that the person experiencing them does not understand. This mechanism is also involved in anxiety disorders where stimuli trigger uncontrolled and disproportionate fear responses (LeDoux 1996, pp. 239–242).

The HCogAff architecture (Fig. 3.3b) shows how losing control can result from primary and secondary emotions (Sloman 2000, 2002). Primary emotions, such as being startled, terrified, delighted, or the thalamo-amygdal fear mechanism described by LeDoux, can be implemented as global interrupts to processing in the lower reactive layer of the HCogAff architecture. Secondary emotions, such as being anxious, apprehensive or relieved, arise from interruptions to deliberative processing in the middle layer. The changing locus of control between processes in the reactive and deliberative layers of the HCogAff architecture helps explain how automatic and controlled processes compete for control in emotional episodes. The third level of the HCogAff architecture is concerned with managing processes that occur in the lower architectural levels. Disruptions or 'perturbances' to this third level are termed tertiary emotions. These can include lapses in attentional control seen in episodes of extreme anger, grief, or longing when in intense love (Petters et al. 2011). Tertiary emotions arise from disturbances to processing in this higher third reflective layer of cognitive architectures. Sometimes these tertiary emotions are impairments but sometimes they can involve acceleration, redirection or tighter control that avoids a disaster.

Interrupt mechanisms may help overcome likely design limitations on future robots and because of this their use may be unavoidable. As Sloman and Croucher (1981) predict: *"the need to cope with a changing and partly unpredictable world makes it very likely that any intelligent system with multiple motives and limited powers will have emotions."* (Sloman and Croucher 1981, p. 1). However, this does not mean that the loss of control that comes when emotional interrupts occur is good, it is just the consequences of reacting too slowly which is bad.

3.4.2 Grief as a Loss of Attentional Control

Attachment relations start early in development and last throughout our lives (Peters and Waters 2010; Petters et al. 2010). Grief is a response to the loss of someone with whom we are strongly attached. Wright et al. (1996) provides of detailed description of how grief as a loss of attentional control can be explained within the HCogAff architecture. In brief, when someone interacts with another with whom they are attached the processes of circulation, percolation and diffusion described above in Sect. 3.3.2 give rise to a distributed multicomponent attachment structure. So this attached individual will possess information about those they are attached to in their perceptual and belief systems. Many motivators will be formed including those which might activate the goals of proximity to the attachment figure or merely to just think about this person. Some of these will then rise in the hierarchy of dispositional control states. These kinds of motivators will interrelate with other control states in complex ways as they diffuse through the architecture. Grief occurs in response to loss of the attachment figure because this diverse collection of control states cannot be quickly dismantled and will therefore for some time continue to be triggered and gain attention. As Wright et al. (1996) note: *"an attachment structure relating to an individual is a highly distributed collection of information stores and active components embedded in different parts of the architecture and linked to many other potential control states. When an attachment structure concerning individual X exists in an agent, almost any information about X is likely to trigger some internal reaction."* (Wright et al. 1996, p. 3). Grief is therefore a tertiary emotion which can endure for lengthy periods as dispositions within the architecture and may be largely outside of awareness.

3.4.3 Loss of Control as a Detrimental Side-Effect When Emotions Boost Motivation

Although he did not use the terminology of contemporary Cognitive Science, Hume highlighted the deep links between cognitive states-like belief and other thoughts and affective states-like motivation and desire when he proposed that: *"reason alone can never be a motive to any action of the will"* (Hume 1739, p. 413). This quote highlights the truism that motivation and preferences are needed for intelligent thought and action (Sloman 2004). What contemporary approaches can add to Hume's introduction to the importance of motives in reasoning is the concept of an information processing architecture such as the HCogAff architecture within which these motives are processed. Within such an architecture, motivations and preferences do not have to necessarily lead to a loss of control of behaviour or attention.

An ideal in the exercise of will is to gain and maintain control of self and environment rather than lose control as often happens when emotions take over. So an intelligent agent should be able to possess multiple desires and use processes of reasoning to select between options and try and achieve those desires in a controlled and deliberate manner. However, there are at least two exceptions when losing control may be beneficial. As noted above, emotional interrupts may occur to produce quick responses to real-time needs. In addition, humans are opportunistic and can switch at short notice from working towards long-term goals to capture short-term opportunities. In the HCogAff architecture, motivators compete to become active in the deliberative subsystem of the architecture. Sometimes highly beneficial long-term goals could remain active for long periods. However sometimes, especially when potent distractions are present or the variable threshold for motive activation to the deliberative subsystem is low due to fatigue or boredom, beneficial long-term goals can be supplanted by shorter term and less beneficial goals. Ainslie (2001) describes how humans discount the value of future rewards according to a hyperbolic curve rather than an exponential curve that would produce consistent choice over time. For example, someone on a diet may have no intention of eating cream cake tomorrow but put a cream cake where it is only seconds away from them eating it and their preference may change. Emotions partly cause this problem by giving rise to intense short-term desires like greed, but they also provide a solution. Short-term temptations can be combated by longer term emotions like pride or a strong sense of love or duty.

So emotional 'hijackings' do not just have to be short-term interrupts where control is returned after the moment of danger has passed. They can also last for considerably longer durations to increase long-term motivation. Someone who has lost out in some way to another person is more likely to confront them and perhaps gain redress if bolstered with the action readiness and extra motivation provided by being angry. Emotions can involve increased heart rate and preferential blood circulation to the muscles providing increasing action readiness but at the cost of performance in fine-grained control. There is also an information level effect where emotions can lead to stronger commitment to goals. For example, because someone is in an emotional state they may be less likely to give up on their goals or consider evidence, which might suggest taking an alternative course of action.

The public expression of strong emotions which are linked to greater commitment can also have social benefits, but these benefits might be gained by convincingly portraying these emotions without really experiencing them. Sloman (2000) discusses how a teacher may discover that real anger can be used to control a classroom, and learn to become angry. However, even if the real anger provides an overall benefit, this loss of control may still lead to some detrimental side effects. The teacher would be better to express anger convincingly without really being angry with its attendant loss of control.

Passions linked to love, empathy, anger, fear, duty and honour provide motivation and may help overcome distractions like hunger, tiredness and from other less important goals. From an engineering and therapeutic perspective, it is clear that many of the motivational benefits of emotion might be provided by non-emotional alternatives. However, in the context of an architecture biased to short-term opportunities the extra motivation provided by emotions may provide an important function in delivering long-term commitment. Within the HCogAff architecture, emotions may not just be triggered when interruptions are needed to cope with fast moving events but also to halt newly surfacing motivators, which may distract from ongoing long-term goals. The next section will discuss circumstances where loss of control might itself be a benefit rather than merely being a detrimental side effect arising from otherwise beneficial mechanisms.

3.4.4 Loss of Control Which Enables Total Commitment

Humans, more than other animals, can display false emotions. We can also detect such deceptions. Thus, the possibility of an evolutionary arms race between deceiver and deceived. Many commentators have suggested a 'Machiavellian Hypothesis' that 'cognitive arms races' between deceiver and deceived may have been a key driving force in the development of human intelligence (Trivers 1971; Humphrey 1976; Alexander 1990; Rose 1980; Miller 1993, all cited in Pinker 1998) and also that of other non-human primates (Byrne and Whiten 1988; Whiten and Byrne 1997). If detection of sham anger is likely and so someone cannot convincingly pretend to be angry they may fall back on actually being angry. However, real anger is a dangerous tool, which can end badly for the person who has lost control in this way.

Pinker (1998) draws upon the work of a number of researchers (Schelling 1960; Trivers 1971, 1985; Daly and Wilson 1998; Hirshleifer 1987; Frank 1988, all cited in Pinker 1998) who have all independently proposed an emotional mechanism which Pinker terms the 'Doomsday Machine'. This label is adopted by Pinker from the film Dr. Strangelove. In this film, the Russians have developed a network of underground bombs with the potential to kill all life on earth. These bombs will be set off automatically in the event of an attack by another country or an attempt to disarm it. The essential property of the Doomsday Machine is that once set up and turned on there is no going back. It makes the Russians immune to threats and blackmail. Pinker quotes the Dr. Strangelove character in this film to emphasise how voluntarily losing options is the essence of the Doomsday machine:

> *"But," Muffley said, "is it really possible for it to be triggered automatically and at the same time impossible to untrigger?"*

> *...Doctor Strangelove said quickly, "But precisely. Mister President, it is not only possible, it is essential. That is the whole idea of this machine. Deterrence is the art of producing in the enemy the fear to attack. And so because of the automated and irrevocable decision making process which rules out human meddling, the Doomsday Machine is terrifying, simple to understand, and completely credible and convincing."* (quoted in Pinker 1998, p. 408)

Pinker shows that Dr. Strangelove's description of how the Doomsday Machine operates in Nuclear Deterrence Strategy can be transferred to human interpersonal conflict. In fact, we can use exactly the same words as Dr. Strangelove in describing the effects of the kinds of extreme and uncontrolled rage possessed by individuals who have committed work place massacres in the USA or who run Amok in Indochinese cultural contexts. The triggering of such strong emotions allows such individuals to possess responses which can be 'irrevocable', 'terrifying', and 'simple to understand'. The threat from such individuals is also 'credible and convincing' as long as their emotional state can be distinguished from sham-emotion pretence. Examples of extreme rage are not always entirely non-cognitive processes. As Pinker (1998, p. 364) describes, such rampages are preceded by lengthy brooding over failure and involve aspects of planning. They are better interpreted as cognitive strategies where compliance to their commitments has been locked in by strong emotions. However, Pinker goes further and asks whether emotions in general allow individuals to get their own way and enforce their will over others in everyday situations. Pinker (1998) presents a strong view of the intimate interrelation between emotion and reason:

> "People consumed by pride, love, rage have lost control. They may be irrational. They may act against their interests. They may be deaf to appeals. [...] But though this be madness there be method in it. Precisely these sacrifices of will and reason are effective tactics in the countless bargains, promises, and threats that make up our social relations. [...] The passions are no vestige of an animal past, no wellspring of creativity, no enemy of intelligence. The intellect is designed to relinquish control to the passions so that they may serve as guarantors of its offers, promises, threats against suspicions that they are lowballs, double-crosses and bluffs. The apparent firewall between passion and reason is not an ineluctable part of the architecture of the brain; it has been programmed in deliberately, because only if the passions are in control can they be credible guarantors." (Pinker 1998, pp. 412–413)

This a quite a grim view of human relations. Pinker is suggesting that people use their emotions to implicitly threaten others. These threats are different from merely threatening actions like hitting or shouting at people because it is the lack of control that is the real danger. In Sect. 3.4.2 processes such as circulation, percolation and diffusion were invoked in explaining how attachments form and ultimately result in the loss of control of attentive processes experienced in grief. These same processes can also account for how individuals build up highly distributed collections of motivators that can give rise to Doomsday Machine emotional responses. Over time motivators for gaining revenge, keeping face, or maintaining honour and social status can become embedded in different parts of the architecture. Through diffusion these motivators become linked to and interrelate with many other potential control states in complex ways. So when a moment comes that they are triggered the individual experiencing these active motives will struggle to act in their current best interests if this involves controlling and subduing a large network of other interlinked motivators. If Doomday Machine emotions remain untriggered then the individual may benefit. Or an individual that merely threatens Doomsday responses may bring about adverse reactions and avoidance by others.

3.4.5 Loss of Control: Due to Blocking of Access to Painful Information

The concept of repression was developed by Freud to explain behavioural patterns observed in his clinical practice in Vienna in the 1890s and early years of the Twentieth Century (Freud 1925–1995; Storr 1989). Freud's practice included a large number of individuals complaining of a phenomenon which came to be known as Conversion Hysteria. The reminiscences of hysterics were notable in two regard: they were not easily accessible to conscious recall; and they were often painful, shameful or alarming (Storr 1989). As Freud described:

> *"How had it come about that the patients had forgotten so many of the facts of their external and internal lives but could nevertheless recollect them if a particular technique was applied? Observation supplied an exhaustive answer to these questions. Everything that had been forgotten had in some way or other been distressing; it had been either alarming or painful or shameful by the standards of the subject's personality. It was impossible not to conclude that that was precisely why it had been forgotten—that is, why it had not remained conscious. In order to make it conscious again in spite of this, it was necessary to overcome something that fought against one in the patient; it was necessary to make efforts on one's own part so as to urge and compel him to remember. The amount of effort required of the physician varied in different cases; it increased in direct proportion to the difficulty of what had to be remembered. The expenditure of force on the part of the physician was evidently the measure of a **resistance** on the part of the patient. It was only necessary to translate into words what I myself had observed, and I was in possession of the theory of **repression**."* (Freud 1925–1995, p. 18)

Over many hours of observation Freud came to see the mind as involving conflict between conscious thoughts and unconscious emotions 'trying' to become conscious and be discharged, like a boil trying to reach the surface of the skin and release its toxins (Storr 1989). However, unlike in the case of a boil, Freud considered that unconscious psychically toxic thoughts were held below consciousness by an active process of defensive control which he termed repression.

For Freud, the main function of repression is minimisation of psychic pain and distress. More recently, from the perspective of contemporary Evolutionary Psychology, Nesse and Lloyd (1992) have proposed that the capacity for repression is an evolutionary adaptation. According to Nesse and Lloyd (1992), some of the adaptive functions which repression may provide include:

- Controlling mental pain in a similar way that endorphins control physical pain. Although sensing pain is usually a useful evolved capacity as the cause of the pain can be removed or avoided, if no actions can be taken to reduce pain and it serves no other continued purpose (such as learning or avoidance) then removing it from awareness may be adaptive as it removes a source of distraction from ongoing thought.
- Inhibiting conscious recognition of socially unacceptable impulses. These impulses may not be intended to be carried out but would rather involve fantasising. Again, removing them from consciousness would decrease anxiety and may remove a source of distraction.

- Repressing the thought of a friend's selfish motives so that a friendship can be maintained. Nesse and Lloyd (1992) describe how certain schizophrenics have been reported to possess an uncanny ability to apprehend secret and unsavoury motives in others. Nesse and Lloyd speculate that this ability may be due to interference with the normal ability to adaptively deceive oneself about the motives of others. This toleration of selfish motives in one's friends may be adaptive or maladaptive but certainly reduces exposure to painful thoughts.

Nesse and Lloyd also propose that the self-deceptive nature of repression allows people to follow their own selfish motives with less chance of detection by others. This fourth proposal for an adaptive function for repression is less about psychic defense and was inspired by ideas presented in Trivers (1976). The next section presents a detailed account of Triver's updated views on possible functions for self-deception (von Hippel and Trivers 2011).

Nesse and Lloyd's attempt to link repression with contemporary Evolutionary Psychology is weakened by claims that repression is not a human universal but arises in particular 'rule-bound' cultural environments. The type of case upon which Freud's theory was originally based—severe conversion hysteria—are seldom observed today. This has been explained by the fact that Freud undertook his studies with subjects from a particular cultural and historical milieu—the Viennese upper or upper middle class culture of the 1890s and early twentieth century. Reasons given for the drop in frequency of hysteria include societal emancipation from the constrictions of Victorian culture and changes in emotional literacy in the twentieth century. These changes would be expected to decrease the 'need' for repression (Shorter 1993; Stone et al. 2008). Stone et al. (2008) provides a contrasting view that suggests hysteria has not disappeared but rather interest has waned in it by clinicians and patients.

Freud lacked an appropriate theory of information processing and control mechanisms but contemporary frameworks such as the HCogAff architecture may be able to capture the spirit of Freud's conceptual ideas. As Fig. 3.3b shows the HCogAff architecture can include a number of 'personae', which are high-level culturally determined templates which cause global features of the behaviour to change, e.g. switching between bullying and servile behaviour (Sloman 2001b). Sloman (2010) argues that personae are required because meta-management subsystems may need different monitoring and control regimes for use in different contexts. Personae might therefore be a mechanism whereby the repression of painful thoughts could be implemented in the HCogAff architecture. Freud noted that repressed thoughts take effort to retrieve. This effort may be the personae being switched, merged or given transparent access to each other.

3.4.6 Loss of Control: Due to Biases in What Information is Processed

Freud's and Nesse and Lloyd's approaches to self-deception are mostly defensive, invoking repression as a process which removes painful thoughts from conscious awareness. In contrast, von Hippel and Trivers (2011) present a more offensive function for self-deception. The central claim of von Hippel and Trivers (2011) is that *"self deception evolved to facilitate interpersonal deception by allowing people to avoid the cues to conscious deception that might reveal deceptive intent"* (von Hippel and Trivers 2011, p. 1). This idea was first presented by Trivers in a foreword to 'The Selfish Gene' (Dawkins 1976):

> *"If (as Dawkins argues) deceit is fundamental in animal communication, then there must be strong selection to spot deception and this ought, in turn, select for a degree of self-deception, rendering some facts and motives unconscious so as to not betray by the subtle signs of self-knowledge the deception being practiced. Thus, the conventional view that natural selection favours nervous systems which produce ever more accurate images of the world must be a very naive view of mental evolution."* (Trivers 1976, pp. 19–20).

In their recent treatment, von Hippel and Trivers (2011) expand on Triver's (1976) conjecture and suggest self-deception is a particularly good way to deceive others because it eliminates the costly cognitive load that can be involved in carrying out deception of others, and it also minimises retribution. Having lots of incorrect beliefs about the world is not beneficial but von Hippel and Trivers suggest that when a person deceives themselves by bolstering their own positive qualities and other peoples negative qualities this can lead to greater confidence and hence social and material advancement.

Where von Hippel and Trivers (2011) differ from the approaches of Freud and Nesse and Lloyd described above is in the variety of processes they claim can give rise to self-deception. What Freud's original approach and Nesse and Lloyd's more recent approach to explaining repression have in common is that the repressed individual possesses two separate representations of reality with truth preferentially stored in the repressed unconscious component and falsehood preferentially in the conscious mind (von Hippel and Trivers 2011). Von Hippel and Trivers suggest that in addition *"the dissociation between conscious and unconscious memories combines with retrieval-induced forgetting and difficulties distinguishing false memories to enable self-deception by facilitating the presence of deceptive information in conscious memory while retaining accurate information in unconscious memory"* (von Hippel and Trivers 2011, p. 6). They also propose that implicit and explicit attitudes dissociate which enables people to express to others socially desirable attitudes whilst simultaneously acting on socially undesirable attitudes. This is because the socially undesirable attitudes are relatively inaccessible and so help the individual maintain plausible deniability. A further suggested mechanism that helps implement self-deception is a dissociation between automatic and controlled processes. So according to von Hippel and Trivers, whilst controlled processes pursue goals that an individual has deceived themselves that they want to pursue, automatic processes pursue the true but hidden goals. As von Hippel and Trivers note: by *"by causing neurologically intact individuals to split some aspects of their self off from others, these dissociations ensure that people have limited conscious access to the contents of their own mind and to the motives that drive their behaviour. In this manner, the mind circumvents the paradox of being both deceiver and deceived."* (von Hippel and Trivers 2011, p. 6)

The processing biases which von Hippel and Trivers (2011, pp. 7–11) detail as bringing about a state of self-deception include: (1) biased information search (which can vary in amount of searching; selection of searching and selective attention); (2) biased interpretation; (3) misremembering; (4) rationalisation; and (5) convincing the self that the lie is true. These biases in processing can be situated within the HCogAff architecture. For example, the distinction between reactive and deliberative processes might be related to the distinctions von Hippel and Trivers (2011) discuss between conscious and unconscious memories, implicit and explicit attitudes, and controlled and automatic processes. In addition, processing biases such as biases in information search, biased interpretation and misremembering might all arise from tertiary emotions where meta-management processes are perturbed. In addition to explaining how these biases can form, the HCogAff architecture should also attempt to explain why meta-management processes do not uncover self-deception.

3.5 Conclusion

This paper has shown that the explanatory framework offered by the CogAff schema and HCogaff architecture provides a promising opportunity for producing deep emotion models from an integrative analysis of different types of loss of control. In particular, primary emotions such as fear or happy surprise and secondary emotions such as worrying about future events can be explained as interrupts to processing in the HCogAff architecture. Processes which describe how motivators relocate and transform within the HCogAff architecture, such as circulation, percolation and diffusion can be used to explain a range of tertiary emotions such as grief and uncontrolled rage. Self-deception involves losing access to 'true' information and how this occurs may spur theoretical developments in the HCogAff architecture. Future work may also not only deepen and enrich a HCogAff explanation of the phenomena discussed above but allow other complex emotional phenomena to be integrated within this explanatory framework.

References

Ainslie G (2001) Breakdown of Will. Cambridge University Press, Cambridge
Alexander RD (1990) How did humans evolve. Reflections on the uniquely unique species. Special Publication No. 1, Museum of Zoology, University of Michigan

Arbib M, Fellous JM (2004) Emotions: from brain to robot. Trends Cogn Sci 8:554–561

Byrne R, Whiten A (1988) Machiavellian intelligence: social expertise and the evolution of intellect in monkeys, apes and humans. Oxford University Press, Oxford

Colombetti G, Thompson E (2008) The feeling body: towards an enactive approach to emotion. In: Overton WF, Muller U, Newman JL (eds) Developmental perspectives on embodiment and consciousness. Lawrence Erlbaum Ass, New York, pp 45–68

Daly M, Wilson M (1988) Homicide. Hawthorne, New York

Darwin C (1872) The expression of the emotions in man and animals. Harper Collins, London (Reprinted 1998)

Dawkins R (1976) The sefish gene. Oxford University Press, Oxford

Ekman P (2003) Emotions revealed. Times Books, New York

Frank RH (1988) Passions within reason: the strategic role of the emotions. Norton, New York

Freud S (1925l1995) An autobiographical study. In: Gay P (ed) The Freud Reader, Vintage, London

Hirshleifer J (1987) On the emotions as guarantors of threats and promises. In: Depre J (ed) The latest on the best: essays on evolution and optimality. MIT Press, Cambridge, Mass

Hume D (1739) A treatise of human nature . Norton DF, Norton MJ (eds) Oxford, Clarendon Press (a critical edn, 2007)

Humphrey NK (1976) The social function of intellect. In: Bateson PG, Hinde RA (eds) Growing points in ethology. Cambridge University Press, London

Kagan J (2007) What is emotion?. Yale University Press, New Haven

LeDoux J (1996) The emotional brain. Simon and Schuster, New York

Miller GF (1993) Evolution of the human brain through runaway sexual selection: the mind as a protean courtship device. Ph.D dissertation, Department of Psychology, Stanford University

Nesse R, Lloyd A (1992) The evolution of psychodynamic mechanisms. In: Barkow JH, Cosmides L, Tooby J (eds) The adapted mind: evolution psychology and the generation of culture. OUP, Oxford, pp 601–624

Petters D (2006) Designing agents to understand infants. Ph.D. thesis, School of Computer Science, The University of Birmingham (Available online at http://www.cs.bham.ac.uk/research/cogaff/)

Petters D, Waters E (2010) A.I., attachment theory, and simulating secure base behaviour: Dr. Bowlby meet the reverend bayes. In: Proceedings of the international symposium on 'AI-inspired biology', AISB Convention 2010. AISB Press, University of Sussex, Brighton, pp 51–58

Petters D, Waters E, Schönbrodt F (2010) Strange carers: robots as attachment figures and aids to parenting. Interact Stud Soc Behav Commun Biol Artif Syst 11(2):246–252

Petters D, Waters E, Sloman A (2011) Modelling Machines which can love: from Bowlby's attachment control system to requirements for romantic robots. Emot Res 26(2):5–7

Pinker S (1998) How the mind works. Penguin Books, London

Rose M (1980) The mental arms race amplifier. Human Ecol 8:285–293

Schelling T (1960) The strategy of conflict. Harvard University Press

Scherer K (2010) Emotion and emotional competence: conceptual and theoretical issues for modelling modelling agents. In: Scherer K, Banziger T, Roesch E (eds) Blueprint for affective computing: a sourcebook. OUP, Oxford, pp 3–20

Shorter E (1993) From paralysis to fatigue: history of psychosomatic illness in the modern era. Simon and Schuster, London

Simon HA (1967) Motivational and emotional controls of cognition. Psychol Rev 74(1):29–39

Sloman A (1995) What sort of control system is able to have a personality? In: Presented at workshop on designing personalities for synthetic actors. Vienna, June 1995

Sloman A (2000) Architectural requirements for human-like agents both natural and artificial. (what sorts of machines can love?). In: Dautenhahn K (ed) Human cognition and social agent technology, advances in consciousness research. John Benjamins, Amsterdam, pp 163–195

Sloman A (2001a) Beyond shallow models of emotion. Cogn Process Int Q Cogn Sci 2(1):177–198

Sloman A (2001b) Evolvable biologically plausible visual architectures. In: Cootes T, Taylor C (eds) Proceedings of British machine vision conference, BMVA, Manchester, pp 313–322

Sloman A (2002) How many separately evolved emotional beasties live within us? In: Trappl R, Petta P, Payr S (eds) Emotions in humans and artifacts. MIT Press, Cambridge, pp 29–96

Sloman A (2004) Simulating infant-carer relationship dynamics. In: Proceedings of AAAI spring symposium 2004: architectures for modeling emotion—cross-disciplinary foundations (No. SS-04-02 in AAAI Technical reports). Menlo Park, pp 128–134

Sloman A (2010) An alternative to working on machine consciousness. Int J Mach Consciou 2(1): 1–18

Sloman A, Chrisley RL (2003) Virtual machines and consciousness. J Conscious Stud 10(4–5): 113–172

Sloman A, Chrisley RL (2005) More things than are dreamt of in your biology: information-processing in biologically-inspired robots. Cogn Syst Res 6(2):145–174

Sloman A, Croucher M (1981) Why robots will have emotions. In: Proceedings of 7th international joint conference on AI. Vancouver, pp 197–202

Stone J, Hewett R, Carson A, Warlow C, Sharpe M (2008) The 'disappearance' of hysteria: historical mystery or illusion? J Roy Soc Med 101(12–18):1

Storr A (1989) Freud. OUP, Oxford

Trivers R (1971) The evolution of reciprocal altruism. Q Rev Biol 46(1):35–57

Trivers R (1976) Foreword. In: Dawkins R (eds) The selfish gene. Oxford University Press, New York, pp 19–20

Trivers R (1985) Social evolution. Benjamin/Cummings, Menlo Park, CA

von Hippel W, Trivers R (2011) The evolution and psychology of self-deception. Behav Brain Sci 34:1–56

Whiten A, Byrne R (1997) Machiavellian intelligence: evaluations and extensions, vol 2. Cambridge University Press, Cambridge

Wright I (1997) Emotional agents. Ph.D. thesis, School of Computer Science, The University of Birmingham, Birmingham

Wright I, Sloman A, Beaudoin L (1996) Towards a design-based analysis of emotional episodes. Philos Psychiatry Psychol 3(2):101–126

Chapter 4
Acting on the World: Understanding How Agents Use Information to Guide Their Action

Jackie Chappell

4.1 Introduction

I am only interested in everything. —Les Murray, poet

In many respects, writing a reflective essay on a topic within Aaron Sloman's research interests is an easy task: after all, he has so many interests! However, the real difficulty lies in narrowing the selection down to focus on a particular *subset* of his interests. Appropriately, this problem of selection or attention is analogous to that faced by animals living in complex environments. They need to work out which aspects of their complex environments are the most interesting or important (where importance is governed by whether or not it affects the animal's fitness), and then store information about those parts in a structured manner that allows them to use that information to guide their actions. This is the problem I will discuss in this essay.

Here I will attempt briefly to outline the current state of the art in animal cognition, specifically referring to how animals gather and represent information about the world in order to take action on it. Throughout I will use the informal meaning of information as referring to semantic content that is *about* something that actually exists or could exist (Sloman 2011). Since this is such a vast topic, I will only be able to touch briefly on a few of the most active areas of research. I will then outline a few of the many unsolved problems, followed by some suggestions about the most promising approaches for investigating these problems.

I refer in the title to 'agents' because most of what I say could apply equally to artificial agents that have to negotiate and explore complex environments and yet still carry out the actions and tasks that they were programmed to complete. Indeed, as will become clear throughout this essay, I think that there are benefits in considering the problem from the perspective of the challenges posed by the environment (which

J. Chappell (✉)
School of Biosciences, University of Birmingham, Edgbaston, Birmingham B15 2TT, UK
e-mail: j.m.chappell@bham.ac.uk

J. L. Wyatt et al. (eds.), *From Animals to Robots and Back: Reflections on Hard Problems in the Study of Cognition*, Cognitive Systems Monographs 22, DOI: 10.1007/978-3-319-06614-1_4, © Springer International Publishing Switzerland 2014

is a more familiar perspective for AI researchers), rather than thinking only about the capabilities of the animal. However, since my own expertise is in animal cognition and not AI, my examples will be drawn from the animal literature, including work on humans.

4.2 What We Know About Animal Cognition

When we observe animals, what often impresses us most is how easily they adapt their behaviour appropriately to rapidly changing contexts, deal with complex environments and solve problems. For example, imagine a Sumatran orangutan (*Pongo abelii*) moving around in the canopy to forage. Orangutans are the largest arboreal animals, and so face severe problems in navigating the discontinuous rainforest canopy. They need to cross gaps between trees, but to do so, they must pass through the zone of thin twigs and branches at the periphery of the trees' canopies, rather than paying the high energetic cost of descending to the ground, crossing and ascending the next tree (Thorpe et al. 2007). These twigs are insufficient to support their weight (males weigh up to 100 kg and females up to 40 kg), so they need to find a way of traversing the gap safely. One way in which orangutans achieve this is to use their body mass to deform compliant branches in order to bridge the gap, including swaying compliant trunks of trees like an inverted pendulum (Thorpe and Crompton 2006, 2005).

Many questions are raised by this kind of behaviour: do orangutans follow habitual routes, or do they plan their route two or more steps in advance to avoid moving into a tree from which it is difficult to cross to another without backtracking? How do they decide what would make a suitable support for a particular crossing? Many different variables are important in determining whether a particular support is suitable, including the size of the gap, the nature of the supports on the far side of the gap, the orangutan's body weight and limb span, the type of locomotion it chooses to use, the type of tree or liana used and its diameter. How does the orangutan collect, store, organize and then use these kinds of information to guide its action? These are real problems that have an important influence on the orangutan's evolutionary fitness: it needs to move through the canopy to feed, and if it falls from the height of the canopy it is likely to be seriously injured.

In the remainder of this section, I will outline what we currently know about these kinds of processes in animals in a selection of domains, before moving on to discuss the areas in which we still lack understanding in Sect. 4.3.

4.2.1 Behavioural Flexibility

One of the features that makes the behaviour of some animals so impressive is the behavioural flexibility they display. While some animals show stereotyped sequences of behaviours irrespective of context, most animals—even some invertebrates—are able to alter their actions depending on the context, or if a sequence of actions is interrupted. In all but the simplest of environments, this kind of capability is necessary to ensure that the animal is able to deal with environments that are highly variable, and that may alter dynamically during execution of behaviours.

Web construction by spiders appears to be a behaviour under fairly tight genetic control. The sequence of actions required in order to build a web are highly constrained, and vary systematically between species. Most orb-weaving spiders show variation in their behaviours while they are attaching the 'proto-hub' to the surrounding supports in the environment. However, once they have built this scaffolding, they can move on their own threads exclusively, and as a consequence the sequence of behaviours used to construct the frame, radii and add the sticky spiral is highly stereotyped (Benjamin and Zschokke 2004). In contrast two species of spiders in the genus *Linyphia* that construct sheet webs show highly variable web construction sequences, frequently switching between building the supporting structure and laying down sticky threads (Benjamin and Zschokke 2004). Furthermore, many species are able to adjust the parameters of their web designs depending upon the prevailing conditions. For example, the spacing between the spirals of orb webs is crucial in determining the size of prey that can be trapped (Sandoval 1994), and spiders can alter the location, orientation and strength of their web depending upon the direction and strength of the wind or other environmental factors (Eberhard 1971; Moore 1977).

Male satin bower birds (*Ptilonorhynchus violaceus*) construct elaborate 'bowers' to attract females. The bowers consist of two parallel walls of sticks planted in the ground and graduated in height, and are decorated with objects of a particular colour (blue in the case of satin bower birds). Females choose males partially on the basis of their bower, and the symmetry of its walls. It has recently been shown that if one wall is experimentally destroyed (something which apparently happens vary rarely in nature), males preferentially replace the sticks in the destroyed wall, not the intact one Keagy et al. 2011. Furthermore, they show a 'templating' behaviour in which they hold a stick vertically in their bill, moving until they match the length of the held stick with its counterpart in the intact wall. They then turn to the same position on opposite side and place the stick in the corresponding position in the destroyed wall (Keagy et al. 2011). This ability to reorder or reorganize actions depending on the current situation allows the animal to begin to 'debug' problems which occur during execution. Without this ability, animals (and robots) can become trapped in a loop when a condition which would allow them to proceed to the next action is not present.

Another aspect of flexibility is the ability to group features in the environment that share some property or functional aspect, and respond to features in the that group in the same way. For example, animals are able to group items into 'food' and 'non-food' categories, and assign novel items to the correct groupings by using shared characteristics. Furthermore, many animals can use relational rules in order to learn, rather than having to learn specific pairings of stimuli. Honey bees (*Apis mellifera*) can learn to choose the stimulus that is above a reference mark on a given trial, even when the stimuli themselves are novel and change on each trial (Avarguès-Weber et al. 2011; Chittka and Jensen 2011). Similarly, a grey parrot (*Psittacus erithacus*) who learned vocal labels for many properties of objects (colour, shape, material, number, etc.) was able to answer questions about which property was shared or differed among a set of presented objects (Pepperberg 1987).

Many animals, therefore, can adapt their behaviour 'online' to deal with variability or inconsistencies in the environment, or to changes outside the animal's own control that occur after the behaviour has been initiated. Others can group their responses to certain classes of features, even if the precise perceptual details of the items in that class vary. Can animals also group their responses according to the physical and causal properties of objects, and predict the effects of their own actions?

4.2.2 Physical Cognition

The physical world provides another source of complexity and variability. The animal encounters objects, materials and surfaces which differ markedly in their properties and the affordances they provide (*sensu* Gibson 1977). In some circumstances, when the variability in the features of the environment that affect the animal's fitness is low, mechanisms of associative learning may suffice. That is, the animal may learn to associate the perceptual qualities of a particular object with a particular affordance. In AI terms, this would be equivalent to recognizing that a cup can be grasped only if the cup is the same size, shape and colour as the one used during training.[1] However, in more complex situations where the number of configurations of variables relevant to the animal is high (such as the situation described for the orangutan at the beginning of this section), there is a potential for combinatorial explosion (Perlovsky 1998; Bellman 1961) and other forms of learning and cognition will be required.

Since the physical world has certain regularities (the principles of continuity, connectedness, solidity and gravity are a few examples), it is possible that animals are born with basic knowledge about these predictable properties, but later add to and modify these rules after experience with the complexity of the world. This appears to be the case for human children, who have certain expectations about the domain of the physical world (as well as number and agency among others; see Spelke and Kinzler 2007). There has been much less research in this area in non-human animals, but recent studies suggest that atleast apes (Cacchione et al. 2009; Cacchione and Call 2010) and domestic dogs (Kundey et al. 2009) may have basic core concepts in the area of solidity.

Physical cognition also involves understanding the causal properties of objects, and the effects that the individual's own actions will have on the world. In many cases this also requires an understanding of the physical properties of objects (for example, whether an object is rigid enough to act as a tool, or whether an object's shape will allow it to roll along a surface), but it extends to understanding the causal

---

[1] Although animals can generalize somewhat between learned stimuli.

role of the actor and each of the objects involved in achieving the desired outcome. This has been extensively investigated in the domain of tool use, but tool use is just one example of a suite of behaviours involving the kinds of information processing described here. We know that New Caledonian crows (*Corvus moneduloides*)—a tool manufacturing species—can select tools with appropriate properties depending on the context of the task (Chappell and Kacelnik 2002, 2004; Bluff et al. 2010), and can even manufacture an appropriate tool (a hook) using a novel material (metal wire) (Weir et al. 2002). Rooks (*Corvus frugilegus*)—a non-tool using species in the wild— were also able to spontaneously modify and use materials as tools, as well as solving other problems involving physical cognition (Bird and Emery 2009a, b). Thus, the kinds of cognitive abilities required to use and manufacture tools do not appear to be adaptive specializations of tool use itself. Indeed, New Caledonian crows appear to have certain morphological adaptations (an unusually straight bill and substantial overlap in their binocular visual field) that make it easier for them to maintain a stable grip on the tool while being able to visually monitor the tip of the tool (Troscianko et al. 2013). Clearly, cognitive abilities are not the whole story. Nevertheless, using tools seems to impose an additional cognitive load on some species, making physical cognition problems more difficult to solve. When chimpanzees were tested on a trap tube problem (in which the subject has to remove a food item from the apparatus while avoiding moving it over a trap), with and without the requirement to use a tool, the chimpanzees performed significantly better when no tool was required (Seed et al. 2009).

One issue which frequently arises in the literature is whether any non-human animals can solve problems involving "unobservable causes", defined as those "based on structural or functional relationships between objects rather than on perceptually based exemplars" (Penn and Povinelli 2007, p. 107). Gravity is the unobservable cause most frequently discussed, but support and weight are also important (see Povinelli 2011 and Buckner 2013 for opposing positions on this debate). If we return to the orangutan example mentioned earlier, the compliance (or flexibility) of supports is not apparent to the animal until it applies its weight to the support, observes another individual doing so, or perhaps observes the effect of the wind on the trees. Diameter of the support might give the orangutan a rough estimate of its likely compliance, but this is likely to be complicated by the material (supports formed from tree branches and lianas will differ in compliance for a given diameter), by the length of the support and whether it is connected to other supports in a mesh of branches. In these circumstances it is possible that animals have certain 'expectations' about the causal effects of their own actions, which they may 'test' by performing actions, much as human children have been shown to do (e.g. Cook et al. 2011; Bonawitz et al. 2012). In contrast, it is clear that great apes are able to use another 'invisible' cue (weight) to infer the location of hidden food (Hanus and Call 2008; Schrauf and Call 2011). Indeed, chimpanzees have been shown to find it easier to use a causal cue (weight) rather than an arbitrary cue (colour) to select a full bottle of juice to open from an array of four other empty bottles (Hanus and Call 2011).

A related question is whether any non-human animals might choose to take actions which do not necessarily bring the subject closer to attaining the goal ('pragmatic actions'), but might provide new information that was formerly hidden, or is difficult to compute mentally ('epistemic actions'). Experiments involving humans playing the computer game 'Tetris' have shown that players sometimes take actions that do not advance the game state towards the goal directly, but reduce cognitive load or mental computation in some way (Kirsh and Maglio 1994). For example, players often rotated Tetris shapes very early in the game, before they could possibly determine the correct final orientation for the shape, suggesting that they were physically rotating the shape to avoid the time cost and cognitive load of mentally rotating it (Kirsh and Maglio 1994). As far as I am aware, no research has yet investigated the occurrence of epistemic actions in non-human animals. It would require very careful analysis of actions to distinguish erroneous pragmatic actions from epistemic actions, but in principle it should be possible to investigate this phenomenon in non-verbal species.

To summarize, there is increasing evidence that a number of non-human animal species collect and use information about the physical properties or causal effects of objects in their environment, possibly even causes that are not directly observable. However, we are still some way from understanding how this information is structured and represented by the animal, and how animals might bring together several actions in sequence to solve a more complex problem.

4.2.3 Planning

In highly complex environments with a great deal of variability, there may be many potential actions to be taken in order to achieve a goal, not all of which will be equally effective. In these situations, planning might help the animal to select among the potential actions. Planning is an immensely complicated topic, and in the field of animal cognition, we are only just starting to investigate how (or indeed, *whether*) non-human animals might plan. In the simplest case, the animal might select one action or a sequence of actions in a pre-determined order, which will immediately result in obtaining the goal. There is a decision among options to be made, but only one step is involved (many people would argue about whether this should really be called planning at all: see Sloman 1999 for a discussion). More complex kinds of planning might involve situations in which each option, if chosen, leads to a selection of further options, and so on, along a branching decision tree. If choices made in the initial steps can determine success or failure many steps ahead, the subject needs to plan its actions prospectively. With even a moderate range of options at each step and a moderate number of steps, the number of possible combinations explodes, making it very difficult to evaluate all the options exhaustively.

Planning for most animals (if it occurs at all) probably falls somewhere between those two extremes. One important issue is how animals select among the alternatives, particularly when there are multiple steps, each with their own consequences. Do animals internally simulate the consequences of alternative actions (that is, as a mental rather than a physical process)? That would imply that they have some kind of internal model of the world (either stored from previous experience, or extrapolated on the basis of the kinds of understanding outlined in Sect. 4.2.2) with which to generate such simulations. The problem for behavioural biologists is that this activity is mostly internal and inaccessible to us, so we are faced with difficulties in determining whether it is occurring at all.

One experimental strategy is to pose a problem and allow subjects a period in which they can observe, but not interact with, the apparatus. The rationale is that if subjects formulate a plan for solving the problem during the preview period using internal simulation, they should be faster at completing the task than subjects denied a preview period. Dunbar et al. (2005) employed this design, presenting four types of puzzle box (which differed in difficulty) to chimpanzees, orangutans and human children. Subjects were presented with the puzzle boxes to open in order to obtain a reward, and were either allowed to view the boxes before attempting to open them ('prior view condition') or not ('no prior view condition'). In all species, there was an effect of the prior view condition such that subjects solved the task faster when given a prior view. However, there are problems with this approach: differences in motivation or behavioural style may affect the time taken to solve the task, but have no implications for internal simulation or its absence. In addition, presenting the elements of the task prior to the opportunity to solve it might speed up the solution of the problem without requiring planning, through the action of priming. That is, subjects might evaluate each separate element of the problem (which might facilitate their interactions with those elements), without putting a functional sequence of actions together into a plan.

Another approach is to examine the choices made by subjects at each step, while altering the difficulty of the task, and including some options that are superficially attractive, but eventually lead to failure when chosen over less attractive, but ultimately successful options. For example, Fragaszy et al. (2003) tested capuchin monkeys and chimpanzees on a 2D computerized maze task, in which subjects had to move a cursor through a maze towards a goal. Both species were more successful at completing the mazes than would be expected by chance, suggesting that they could 'look ahead' and plan their path appropriately. However, capuchins were less successful when they had to take a path that initially lead away from the goal direction (Fragaszy et al. 2009). This experimental approach is a useful one, but has the disadvantage that it requires extensive training of subjects to use the apparatus. In an attempt to design a task probing similar processes, but which did not require training or the use of tools, we tested orangutans on a horizontal puzzle tube, in which two sets of obstacles on each side of a centrally placed reward could hinder the subject retrieving the food (Tecwyn et al. 2012). Subjects therefore had to look ahead to determine which of the two alternative paths would allow them to obtain the food. Two of the three subjects solved the task, but there were interesting differences in strategy between them, suggesting the use of different combinations of rules in order to solve the task (Chappell and Hawes 2012).

4.3 Unsolved Problems in Animal Cognition

I have presented the research in Sect. 4.2 as the 'state of the art', but in reality we are still far from having a complete picture of how animals collect, store, represent and use information about the world.[2] When working with fully competent organisms (in contrast to robots or simulations that you construct from scratch), it is easy to take certain aspects of their behaviour for granted. In particular, the more we probe in certain areas, the more we realize that we lack detailed information about the mechanisms involved, or that the diversity of possible cases is much greater than we initially appreciated (see Sect. 4.2.3). In this section, I will summarize a few of the areas where our knowledge is lacking, focussing particularly on the ways in which animals gather information.

4.3.1 Perception

In order to collect information about the world, animals first have to perceive it: that is, they need to receive and translate signals from the environment, then store them as neural representations. It may seem a relatively trivial task to work out what the perceptual content is for animals, particularly as we understand the mechanisms of many animal sensory systems quite well. However, we encounter problems when studying species that have rather different sensory systems from our own. To give one example, many species of birds have laterally placed eyes, in contrast to our own which are frontally placed. Thus, it is difficult to determine what the bird is attending to without having detailed knowledge about the arrangement and extent of both the monocular and binocular visual fields. Since the arrangement of birds' visual fields are dependent upon their ecological niche (Martin 2007), these can differ substantially between species, and their arrangement and coverage can often be surprising.

For example, we have recently shown that while Senegal parrots (*Poicephalus senegalus*) have a relatively broad frontal binocular field and good visual coverage above and somewhat behind the head, they cannot see below their bill tip (Demery et al. 2011). This means that they cannot see what is held in their bill, which has important implications for understanding what information they can collect from their environment during exploration (see Sects. 4.3.2 and 4.3.3). Conversely, (as alluded to earlier), the combination of a broad binocular overlap and a straight bill in New Caledonian crows allows them to both hold a tool securely and observe the tip of the tool while they are using it (Troscianko et al. 2013).

---

[2] Note that while it is easier to think about these processes as sequential, they certainly occur in parallel in nature.

4.3.2 Attention

Animals have limited capacity for processing information at any one time, which tends to be less than the rate at which the environment provides information. Thus, they have to direct their attention selectively. While we know a great deal about the physiological and neurobiological mechanisms of attention (particularly visual attention) in humans (Desimone and Duncan 1995), most of the research in non-human animals have been conducted on other primates or rats. However, it is clear that attention has a major role in shaping the information processed by animals in important areas of their lives (Dukas 2002). For example, when preying on simulated cryptic (camouflaged) prey, located in their central visual field, blue jays were significantly less likely to detect a stimulus in their peripheral visual field (simulating detection of a predator) when the prey detection task was more difficult (Dukas and Kamil 2000), even though their success rate for detecting prey did not differ between the easy and difficult task.

Attention therefore has both perceptual and cognitive aspects: the sense organs need to be directed towards a stimulus, but processing resources also need to be focussed on particular parts of the environment, effectively filtering it. Since attentional mechanisms filter and restrict the information that is available for an animal to learn, attention can act as a kind of feedback loop: learning and genetic predispositions can influence what the animal attends to, and what the animal learns is constrained by what it attends to. It is therefore very important for us to attempt to understand attentional mechanisms in animals to get a fuller picture of their whole information processing system.

How can behavioural biologists go about studying this? In some cases there may be 'behavioural markers' of attention for sensory modalities such as vision (but see Sect. 4.3.1 on the difficulty of determining visual attention in non-human animals), but in others, attention may be difficult or impossible to observe in naturally behaving animals. We may need assistance from physiologists and neuroscientists to identify such behavioural markers, or we could use objects equipped with instruments in order to detect, for example, the location of touch and force generated by the animal during haptic exploration of an object. Collaborations with computer scientists may also help us to model the putative effects of attention and thus identify markers of attention in animals, such as Heinke and Humphreys' Selective Attention for Identification Model (SAIM) (Heinke and Humphreys 2003).

4.3.3 Exploration

Exploration[3] has been studied extensively in non-human animals in relation to navigation and spatial behaviour, but there has been less research on more general aspects

---

[3] That is, interacting with the environment without an immediate goal other than the collection of information.

of exploration and play (Held and Špinka 2011; Power 2000; Chappell et al. 2012). As with attention (see Sect. 4.3.2), we need behavioural markers of exploration, which can be difficult to define. In an excellent collaboration between biologists and roboticists, Grant et al. (2009) made detailed behavioural observations of the way in which rats used their vibrissae to explore their environment, and showed that they alter the speed and pattern of their 'whisking' behaviour in order to learn about the shape of objects and their surface texture. They then confirmed that the pattern of this behaviour optimizes the efficiency with which this information is collected by modelling the behaviour in a robot (Pearson et al. 2007).

Since information collected during exploration provides some of the raw material for animals' information processing, we need to study the process in more detail (Chappell et al. 2012). For example, do animals start out with perceptual or attentional biases towards exploring certain kinds of objects or collecting certain kinds of information? This could both limit and shape the kinds of information processing they are later capable of (as discussed in Sect. 4.3.2). Is exploration directed primarily towards novel stimuli, in order to collect new information? Do any animals 'test' the properties of objects when their expectations are violated, or reinitiate further bouts of exploration, in the way that we know happens in human children (Cook et al. 2011; Bonawitz et al. 2012)?

4.4 Where Do We Go Grom Here?

In the preceding sections, I have attempted to provide a brief summary of the state of our knowledge about animal cognition as it relates to the physical domain, and have highlighted areas where our knowledge is currently lacking for various reasons. This is, of course, a huge and complex topic, and since most of the interesting stuff goes on 'invisibly' inside the animal's brain, it can be a challenge to investigate. What, therefore, is the best way to proceed?

We still have a long way to go before we have established what mechanisms are involved in the kinds of processes I have discussed above. For example, despite some elegant experiments we have little idea for many species about the different kinds of processes (in the world) that the animal can distinguish between. We also know little about how animals organize the information they collect, whether they form abstractions of some kind, and how they reuse information at a later time or in a different context. Well-designed experiments can probe some of these issues by examining the details of the ways in which animals choose between different options, how they approach problems and so on. We can also present animals with objects which have unusual properties or behave deceptively, and use animals' initial actions and subsequent reactions to those objects to investigate whether they have prior expectations, using techniques similar to those used with pre-verbal human infants.

However, this kind of experimental approach would be much more fruitful if combined with a requirements analysis (Chappell and Thorpe 2010; Sloman 2005).

This involves identifying what kinds of problems and constraints the environment (including other individuals of the same or other species) impose upon the animal and the functions it will need to perform, and thus the kinds of 'designs' of information processing systems that might fulfil those requirements. If we understand what the requirements of the environment are (requirements will differ substantially between species with differing niches), we will be in a better position to focus our own attention on the relevant questions, and avoid overlooking important possibilities because the behaviour of the animal is too subtle to be noticed. This brings me back again to the importance of understanding the sensory and attentional world of the animal being studied: as a strongly visual species, it is easy for us to overlook information in other modalities which might be unavailable or barely perceptible to our own senses.

We also need to examine the precise behaviours of individuals in much more detail, paying particular attention to the errors that they make, as these are likely to be revealing of the operation of their information processing systems, much as bugs in software can reveal important features in the structure of the underlying code. If I have learned anything while studying animal cognition, it is that there is enormous variation between individuals. Thus, while the mean performance of subjects is interesting in some respects, in the face of high variation within a population it can be relatively uninformative about the processes involved. The genetic and developmental histories of individuals (Chappell and Sloman 2007), as well as their experiences throughout life, can have a significant effect on the range and extent of their capabilities (Sloman and Chappell 2007). It is a complex knot to unpick, but attempting to understand how and why individuals differ in their cognitive capabilities might lead us to important insights about the processes and mechanisms involved.

Finally, it is my opinion that we can gain enormous advantages from working with roboticists and other computer scientists. AI can inspire biology as effectively as the reverse position (Chappell and Thorpe 2010): having to construct working robots reveals gaps in knowledge and flaws in assumptions quickly, and robots and simulations can allow you to test hypotheses which would be impossible (for practical or ethical reasons) to test in animals (Grant et al. 2009; Webb 2000). Even if you do not build a working robot or simulation, the exercise of thinking how one might go about it can provide a valuable thought experiment.

These are deep and highly complex problems, and the more closely one studies the problem, the more detail and complexity is revealed, like zooming into an image of a fractal, but without the conceptual benefit of knowing the equation that describes the structure. It is certainly true that the more closely researchers have studied the capabilities of animals, the more advanced capabilities they have found. The bridge between 'uniquely human' abilities and those of other animals remains, but is being gradually eroded as we discover more ways in which behaviour can differ. We cannot solve these problems as isolated disciplines: I hope that we can pool our expertise and continue to be "only interested in everything".

Acknowledgments First, I would like to acknowledge my deep gratitude to Aaron Sloman for many fascinating and stimulating discussions about information processing, evolution and exploration (among many other topics). These conversations have helped me helped me to approach these problems in a new and more productive way. I would also like to thank Nick Hawes, Zoe Demery and Emma Tecwyn for productive discussions on these topics.

References

Avarguès-Weber A, Dyer AG, Giurfa M (2011) Conceptualization of above and below relationships by an insect. Proc Royal Soc B Biol Sci 278(1707):898–905

Bellman R (1961) Adaptive control processes: a guided tour. Princeton University Press, Princeton, NJ

Benjamin SP, Zschokke S (2004) Homology, behaviour and spider webs: web construction behaviour of Linyphia hortensis and L. triangularis (Araneae: Linyphiidae) and its evolutionary significance. J Evol Biol 17(1):120–130

Bird CD, Emery NJ (2009a), Insightful problem solving and creative tool modification by captive nontool-using rooks. Proc Nat Acad Sci USA 106(25):10,370–10,375

Bird CD, Emery NJ (2009b) Rooks use stones to raise the water level to reach a floating worm. Curr Biol 19:1410–1414

Bluff LA, Troscianko J, Weir AAS, Kacelnik A, Rutz C (2010) Tool use by wild new Caledonian crows Corvus moneduloides at natural foraging sites. Proc R Soc B Biol Sci 277(1686): 1377–1385

Bonawitz E, van Schijndel T, Friel D, Schulz L (2012) Children balance theories and evidence in exploration, explanation, and learning. Cogn Psychol 64(4):215–234

Buckner C (2013) In search of balance: a review of Povinellis world without weight. Biol Philos 28(1):145–152

Cacchione T, Call J (2010) Intuitions about gravity and solidity in great apes: the tubes task. Dev Sci 13(2):320–330

Cacchione T, Call J, Zingg R (2009) Gravity and solidity in four great ape species (Gorilla gorilla, Pongo pygmaeus, Pan troglodytes, Pan paniscus): Vertical and horizontal variations of the table task. J Comp Psychol 123(2):168–180

Chappell J, Hawes N (2012) Biological and artificial cognition: what can we learn about mechanisms by modelling physical cognition problems using artificial intelligence planning techniques? Philos Trans R Soc Lond B Biol Sci 367(1603):2723–2732

Chappell J, Kacelnik A (2002) Tool selectivity in a non-primate, the new Caledonian crow (Corvus moneduloides). Anim Cogn 5(2):71–78

Chappell J, Kacelnik A (2004) Selection of tool diameter by new Caledonian crows Corvus moneduloides. Anim Cogn 7(2):121–127

Chappell J, Sloman A (2007) Natural and artificial meta-configured altricial information-processing systems. Int J Unconventional Comput 3(3):211–239

Chappell J, Thorpe S (2010) AI-inspired biology: does AI have something to contribute to biology? In: Proceedings of the international symposium on AI inspired biology: a symposium at the AISB 2010 Convention, Leicester, UK, SSAISB.

Chappell J, Demery ZP, Arriola-Rios V, Sloman A (2012) How to build an information gathering and processing system: lessons from naturally and artificially intelligent systems. Behav Process 89(2):179–186

Chittka L, Jensen K (2011) Animal cognition: concepts from apes to bees. Curr Biol 21(3): R116–119

Cook C, Goodman ND, Schulz LE (2011) Where science starts: spontaneous experiments in preschoolers' exploratory play. Cognition 120(3):341–349

Demery ZP, Chappell J, Martin GR (2011) Vision, touch and object manipulation in Senegal parrots Poicephalus senegalus. Proce R Soci B Biol Sci 278:3687–3693

Desimone R, Duncan J (1995) Neural mechanisms of selective visual attention. Ann Rev Neurosci 18:193–222

Dukas R (2002) Behavioural and ecological consequences of limited attention. Philos Trans R Soc Lond B Biol Sci 357(1427):1539–1547

Dukas R, Kamil AC (2000) The cost of limited attention in blue jays. Behav Ecol 11(5):502–506

Dunbar RIM, McAdam MR, O'connell S (2005) Mental rehearsal in great apes (Pan troglodytes and Pongo pygmaeus) and children. Behav Process 69(3):323–330

Eberhard WG (1971) The ecology of the web of Uloborus diversus (Araneae: Uloboridae). Oecologia 6(4):328–342

Fragaszy D, Johnson-Pynn J, Hirsh E, Brakke K (2003) Strategic navigation of two-dimensional alley mazes: comparing capuchin monkeys and chimpanzees. Anim Cogn 6(3):149–160

Fragaszy DM, Kennedy E, Murnane A, Menzel C, Brewer G, Johnson-Pynn J, Hopkins W (2009) Navigating two-dimensional mazes: chimpanzees (Pan troglodytes) and capuchins (Cebus apella sp.) profit from experience differently. Anim Cogn 12(3):491–504

Gibson JJ (1977) The theory of affordances. The ecological approach to visual perception. Lawrence Erlbaum Associates, London, pp 127–143

Grant RA, Mitchinson B, Fox CW, Prescott TJ (2009) Active touch sensing in the rat: anticipatory and regulatory control of whisker movements during surface exploration. J Neurophysiol 101(2):862–874

Hanus D, Call J (2008) Chimpanzees infer the location of a reward on the basis of the effect of its weight. Curr Biol 18(9):R370–R372

Hanus D, Call J (2011) Chimpanzee problem-solving: contrasting the use of causal and arbitrary cues. Anim cogn 14(6):871–878

Heinke D, Humphreys GW (2003) Attention, spatial representation, and visual neglect: simulating emergent attention and spatial memory in the selective attention for identification model (saim). Psychol Rev 110(1):29–87

Held SD, Špinka M (2011) Animal play and animal welfare. Anim Behav 81(5):891–899

Keagy J, Savard JF, Borgia G (2011) Complex relationship between multiple measures of cognitive ability and male mating success in satin bowerbirds, Ptilonorhynchus violaceus. Anim Behav 81(5):1063–1070

Kirsh D, Maglio P (1994) On distinguishing epistemic from pragmatic action. Cogn Sci 18(4): 513–549, doi:10.1207/s15516709cog1804_1. http://dx.doi.org/10.1207/s15516709cog1804_1

Kundey SMA, de Los Reyes A (2009) Domesticated dogs' (Canis familiaris) use of the solidity principle. Anim Cogn 13(3):497–505

Martin GR (2007) Visual fields and their functions in birds. J Ornithol 148(S2):547–562

Moore CW (1977) The life cycle, habitat and variation in selected web parameters in the spider, Nephila clavipes Koch (Araneidae). Am Midl Nat 98(1):95–108

Pearson MJ, Pipe AG, Melhuish C, Mitchinson B, Prescott TJ (2007) Whiskerbot: a robotic active touch system modeled on the Rat Whisker sensory system. Adapt Behav 15(3):223–240

Penn DC, Povinelli DJ (2007) Causal cognition in human and nonhuman animals: a comparative, critical review. Ann Rev Psychol 58:97–118

Pepperberg IM (1987) Acquisition of the same different concept by an African gray parrot (Psittacus erithacus)—learning with respect to categories of color, shape, and material. Learn Behav 15(4):423–432

Perlovsky L (1998) Conundrum of combinatorial complexity. IEEE Trans Pattern Anal Mach Intell 20(6):666–670

Povinelli D (2011) World without weight: perspectives on an alien mind. Oxford University Press, Oxford

Power TG (2000) Play and exploration in children and animals. Lawrence Erlbaum Associates, London

Sandoval C (1994) Plasticity in web design in the spider Parawixia bistriata: a response to variable prey type. Funct Ecol 8(6):701–707

Schrauf C, Call J (2011) Great apes use weight as a cue to find hidden food. Am J Primatol 73(4):323–334

Seed AM, Call J, Emery NJ, Clayton NS (2009) Chimpanzees solve the trap problem when the confound of tool-use is removed. J Exp Psychol Anim Behav Process 35(1):23–34

Sloman A (1999) What sort of architecture is required for a human-like agent. In: Wooldridge M, Rao A (eds) Foundations of rational agency. Kluwer Academic Publishers, Dordrecht, pp 35–52

Sloman A (2005) The design-based approach to the study of mind (in humans, other animals, and machines), including the study of behaviour involving mental processes. http://cs.bham.ac.uk/research/projects/cogaff/misc/design-based-approach.html

Sloman A (2011) What's information, for an organism or intelligent machine? How can a machine or organism mean? In: Dodig-Crnkovic G, Burgin M (eds) Information and computation. World Scientific, New Jersey, pp 393–438

Sloman A, Chappell J (2007) Computational cognitive epigenetics. Behav Brain Sci 30(4):375–376

Spelke ES, Kinzler KD (2007) Core knowledge. Dev Sci 10(1):89–96

Tecwyn EC, Thorpe SKS, Chappell J (2012) What cognitive strategies do orangutans (Pongo pygmaeus) use to solve a trial-unique puzzle-tube task incorporating multiple obstacles? Anim Cogn 15(1):121–133

Thorpe SKS, Crompton RH (2005) Locomotor ecology of wild orang-utans (Pongo pygmaeus abelii) in the Gunung Leuser ecosystem, Sumatra, Indonesia: A multivariate analysis using log-linear modelling. Am J Phys Anthropol 127:58–78

Thorpe SKS, Crompton RH (2006) Orangutan positional behavior and the nature of Arboreal locomotion in Hominoidea. Am J Phys Anthropol 131:384–401

Thorpe SKS, Crompton RH, Alexander RM (2007) Orangutans utilise compliant branches to lower the energetic cost of locomotion. Biol Lett 3:253–256

Troscianko J, von Bayern AM, Chappell J, Rutz C, Martin GR (2013) Extreme binocular vision and a straight bill facilitate tool use in new caledonian crows. Nat Commun 3:1110

Webb B (2000) What does robotics offer animal behaviour? Anim Behav 60(5):545–558

Weir AAS, Chappell J, Kacelnik A (2002) Shaping of hooks in new Caledonian crows. Science 297(5583):981

Chapter 5
A Proof and Some Representations

Manfred Kerber

Abstract Hilbert defined proofs as derivations from axioms via the modus ponens rule and variable instantiation (this definition has a certain parallel to the 'recognise-act cycle' in artificial intelligence). A pre-defined set of rules is applied to an initial state until a goal state is reached. Although this definition is very powerful and it can be argued that nothing else is needed, the nature of proof turns out to be much more diverse, for instance, changes in representation are frequently done. We will explore some aspects of this by McCarthy's 'mutilated checkerboard' problem and discuss the tension between the complexity and the power of mechanisms for finding proofs.

5.1 Introduction

The idea for the mechanisation of proofs goes back at least to Descartes (1637) and Leibniz's idea to build a reasoning machine by which reasoning can be reduced to computation ("Calculemus!").

The development of logic in the second half of the nineteenth century and the first half of the twentieth century has given the endeavour a great boost. With the arrival of the computer also the right to build such a machine was around and from

This contribution should have been written jointly with Martin Pollet with whom the author worked on this topic in the past. It would have much benefited from this. Many of the thoughts expressed herein are originally Martin's, however, it is hard to trace after a long time in detail who had a certain idea first. Since Martin has moved on and is working on different matters now he did not want to become a co-author and all inaccuracies, inconsistencies, and shortcomings are exclusively the author's. The author would also like to thank Aaron Sloman with whom he had many discussions over many years and which inspired this and related work.

M. Kerber (✉)
Computer Science, University of Birmingham, Birmingham B15 2TT, UK
e-mail: M.Kerber@cs.bham.ac.uk
URL: http://www.cs.bham.ac.uk/mmk

J. L. Wyatt et al. (eds.), *From Animals to Robots and Back: Reflections on Hard Problems in the Study of Cognition*, Cognitive Systems Monographs 22, DOI: 10.1007/978-3-319-06614-1_5, © Springer International Publishing Switzerland 2014

Table 5.1 McCarthy's problem description of 1964, McCarthy (1964)

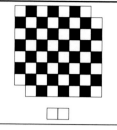

"It is impossible to cover the mutilated checkerboard shown in the figure with dominoes like the one in the figure. Namely, a domino covers a square of each color, but there are 30 black squares and 32 white squares to be covered."

the beginning of artificial intelligence as a field there was the dream to build a reasoning machine. Following Hilbert's programme (and technical advances as those by Herbrand (1930)), the big problem was the instantiation rule for variables since it typically meant that by instantiating a variable x by ground terms (that is, terms without variables in them), e.g. by a, $f(a)$, $f(f(a))$, $f(f(f(a)))$, and so on, infinitely many immediate consequences can be drawn. That is, the search space is infinitely branching. Robinson (1965) solved this problem by the invention of the resolution calculus and the unification algorithm: A method had been invented by which proofs in first order logic can be searched for in a finitely branching search space. Ever since, this calculus and related ones have been refined and efficient implementations (in particular by using indexing techniques inspired from the database field) were developed. In 1996, McCune (1997) proved the Robbins conjecture with his EQP prover, a problem which prominent mathematicians tried to solve but which defied solution for more than 50 years.

However, despite of this success story, the fully automated solution of hard mathematical problems seems more to be the exception than the rule. These systems are more often used in a person-machine tandem. Interactive systems such as Isabelle or Coq allow the proof development guided by a human or a group of humans and fully automated theorem provers are used to fill in smaller gaps in the proof.

A particular problem, the so-called mutilated checkerboard problem, which many people do not consider as trivial when they see it for the first time, can be used to exemplify some of the story. We will look at the problem, its possible formalisations and some solutions in the following.

5.2 The Mutilated Checkerboard Problem

The Mutilated Checkerboard Problem was put as a challenge to the theorem prover community in 1964 by McCarthy (1964).

McCarthy (1964) gives two problem formalisations in his original presentation of the problem. In the first formulations, the problem is represented in a predicate logic language without function symbols and without equality. The signature consists of

Table 5.2 McCarthy's first formalisation of 1964

$S(x, y)$	$y = x + 1$
$L(x, y)$	$x < y$
$E(x, y)$	$x = y$
$G^1(x, y), \ldots, G^4(x, y)$	square (x, y) and the right/top/left/bottom neighbour square are covered by a domino
$G^5(x, y)$	square (x, y) is uncovered

The problem is stated in form of a set of unsatisfiable axioms.

1. $S(1, 2) \wedge S(2, 3) \wedge S(3, 4) \wedge S(4, 5)$
   $\wedge\ S(5, 6) \wedge S(6, 7) \wedge S(7, 8)$ ⎫
2. $S(x, y) \rightarrow L(x, y)$ ⎪ properties of numbers
3. $L(x, y) \wedge L(y, z) \rightarrow L(x, z) \wedge \neg S(x, z)$ ⎬
4. $L(x, y) \rightarrow \neg E(x, y)$ ⎪
5. $E(x, x)$ ⎭
6. $G^1(x, y) \vee G^2(x, y) \vee G^3(x, y) \vee G^4(x, y) \vee G^5(x, y)$ ⎫
7. $G^1(x, y) \rightarrow \neg(G^2(x, y) \vee G^3(x, y) \vee G^4(x, y) \vee G^5(x, y))$ ⎪
8. $G^2(x, y) \rightarrow \neg(G^3(x, y) \vee G^4(x, y) \vee G^5(x, y))$ ⎬ placement of dominoes
9. $G^3(x, y) \rightarrow \neg(G^4(x, y) \vee G^5(x, y))$ ⎪
10. $G^4(x, y) \rightarrow \neg G^5(x, y)$ ⎭
11. $G^5(1, 1) \wedge G^5(8, 8)$ ⎫ uncovered squares
12. $G^5(x, y) \rightarrow (E(1, x) \wedge E(1, y)) \vee (E(8, x) \wedge E(8, y))$ ⎭
13. $S(x_1, x_2) \rightarrow (G^1(x_1, y) \equiv G^3(x_2, y))$ ⎫ adjacency of dominoes
14. $S(y_1, y_2) \rightarrow (G^2(x, y_1) \equiv G^4(x, y_2))$ ⎭
15. $\neg G^3(1, y) \wedge \neg G^1(8, y) \wedge \neg G^2(x, 8) \wedge \neg G^4(x, 1)$ } border of the board

constants $1, \ldots, 8$, and the binary relations given here with their intended meaning. (All variables x, y, and z are implicitly universally quantified.)

Although the problem is finite, it was a challenge in 1964. Today, it is no longer with the vastly enhanced computation power. Still, the proof following the lines of the axiomatisation given in Table 5.2—essentially by checking all cases and ruling them out more or less intelligently—does not reveal a simple *good reason* of why it is impossible, a good reason such as McCarthy himself gives in the exposition of the problem: "*Namely, a domino covers a square of each colour, but there are 30 black squares and 32 white squares to be covered.*"

From a purely logical point of view, a proof just establishes the truth of the theorem and a proof is a proof, whether elegant or not. However, in several ways elegant proofs are better than non-elegant ones. There is first the aesthetic argument, as Hardy put it in (Hardy 1940, p. 29):

> We do not want many 'variations' in the proof of a mathematical theorem: 'enumeration of cases', indeed, is one of the duller forms of mathematical argument. A mathematical proof should resemble a simple and clear-cut constellation, not a scattered cluster in the Milky Way.

Second, clear simple proofs can more easily conveyed and checked. And third, good arguments give a chance for the generalisation of the theorem. From the verbal

Table 5.3 McCarthy's second formalisation of 1964

$1'.\ s(s(s(s(s(s(s(s(8))))))))) = 8$ ⎤
$2'.\ \neg s(s(s(s(x)))) = x$ ⎦ eight distinct numbers

$3'.\ g(x, y) = 5 \equiv x = 8 \wedge y = 8 \vee x = 1 \wedge y = 1$ } uncovered squares

$4'.\ g(x, y) = 1 \equiv g(s(x), y) = 3$ ⎤
$5'.\ g(x, y) = 2 \equiv g(x, s(y)) = 4$ ⎦ adjacency of dominoes

$6'.\ g(1, y) \neq 3 \wedge g(8, y) \neq 1 \wedge g(x, 1) \neq 4 \wedge g(x, 8) \neq 2$ } border of board

$7'.\ 1 = s(8) \wedge 2 = s(1) \wedge 3 = s(2) \wedge 4 = s(3) \wedge 5 = s(4)$ } names

$8'.\ g(x, y) = 1 \vee g(x, y) = 2 \vee g(x, y) = 3 \vee g(x, y) = 4 \vee g(x, y) = 5$ } covering

proof *"Namely, a domino covers a square of each colour, but there are 30 black squares and 32 white squares to be covered."*, it is an easy generalisation to the corresponding theorem *and* proof for the $(2n) \times (2n)$ mutilated checkerboard problem. *"Namely, a domino covers a square of each colour, but there are $2n^2 - 2$ black squares and $2n^2$ white squares to be covered."* With the original representation in first order logic as presented in Table 5.2 it would be very hard to come up with a formalisation of let us say a 1000×1000 mutilated checkerboard. And certainly there is a number n so that checking all possible cases for the $(2n) \times (2n)$ mutilated checkerboard problem is out of reach even for the most powerful computers.

In Sect. 5.3, we will look at several generalisations of the theorem.

5.3 Re-representations of the Problem

In his original paper of 1964, McCarthy gives also a second formalisation, see Table 5.3, in first-order logic, using function symbols (s for successor and g so that $g(x, y)$ is $i \in \{1, 2, 3, 4, 5\}$ according to which of the $G^i(x, y)$ holds), and equality, which is shorter than the first.

An important question—that to a large degree is still unanswered – is how can we get from informal problem descriptions to formal ones. Obviously we want that when the formal problem is solved that this convinces us that the original, informal one is solved as well. This puts the question up what two formalisations have to do with each other, how they can be related and why the solution of the one problem solves the other. Even in this special easy situation the answer is not straightforward. The second formalisation makes use of a cyclic structure (by $s(s(s(s(s(s(s(s(8))))))))) = 8$) and proves a slightly generalised problem, namely that on a torus the mutilated checkerboard problem cannot be solved. Hence, it cannot be solved when the torus is cut and rolled out in a plane, since there are fewer possible coverings in the resulting fragment of the plane. (That is, on the torus there may be coverings which have no correspondence in the plane.)

The transition from a representation of the mutilated checkerboard as a fragment of the two dimensional plane to a fragment of a torus is a slight generalisation.

In general, it is a strong technique used in mathematics to explore a problem by looking for related problems and the attempt to come up with a formulation that is as general as possible. Depending on the representation and the knowledge available to the prover (and/or proof checker), the corresponding proof may be easier in a generalised formalisation. This is true for the mutilated checkerboard problem, and typical for proofs by mathematical induction.

Some simple related problems are mutilated checkerboards with different side length, e.g. 4×4, 3×3, and 2×2.

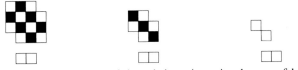

Note that for the case of a 3×3 board there is a simpler proof based on the cardinality of the tiles, but the proof based on there being more white than black tiles works as well. Informative are also cases in which the property does not hold (that is, in this case mutilations to the checkerboard where it is possible to find coverings).

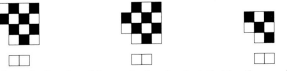

The exploration of related problems, the border that divides the conditions where the theorem holds and where not, gives valuable information to humans when proving a theorem. This can also be a lengthy process and can change the underlying definitions. Certain concepts in mathematics are defined so that the corresponding theorems hold (e.g. $(-1) \cdot (-1)$ is defined as 1 so that the distribution law also holds for negative numbers, $0 = (1 - 1) \cdot (-1) = -1 + 1$). Lakatos' book on Proofs and Refutations Lakatos (1976) exemplifies this with Euler's polyhedron formula.

5.4 Some Proofs

The problem has been restated by McCarthy at the second QED workshop in Warsaw in 1995 as a new challenge to theorem provers in a call for what he called *heavy duty set theory*, see McCarthy (1995). Theorem provers should be able to allow for formalisations which are close to mathematical intuition. That is, it should be possible to tell a theorem prover the trick of colouring the checkerboard in black and white fields so that it becomes possible to speak about the cardinality of black and white fields and that the tiles can cover always only equal numbers of black and white fields.

This challenge was taken up several times, immediately at the QED workshop by Bancerek (1995), who proved the theorem in Mizar using a formalisation which is very close to McCarthy's 1995 formalisation as shown in Table 5.4.

Table 5.4 McCarthy's 1995 Formalisation

Definitions

$$Board = \mathbb{Z}_8 \times \mathbb{Z}_8$$

$$mutilated\text{-}board = Board \setminus \{(0,0),(7,7)\}$$

$$domino\text{-}on\text{-}board(x) \equiv (x \subset Board) \wedge card(x) = 2$$
$$\wedge (\forall x_1 \, x_2)(x_1 \neq x_2 \wedge x_1 \in x \wedge x_2 \in x$$
$$\rightarrow adjacent(x_1,x_2))$$
$$\equiv (x \subset Board) \wedge card(x) = 2$$
$$\wedge (\forall x_1 \, x_2)(x = \{x_1,x_2\} \rightarrow adjacent(x_1,x_2))$$

$$adjacent(x_1,x_2) \equiv |c(x_1,1) - c(x_2,1)| = 1 \wedge c(x_1,2) = c(x_2,2)$$
$$\vee |c(x_1,2) - c(x_2,2)| = 1 \wedge c(x_1,1) = c(x_2,1)$$
$$\equiv |c(x_1,1) - c(x_2,1)| + |c(x_1,2) - c(x_2,2)| = 1$$

$$c((x,y),1) = x$$

$$c((x,y),2) = y$$

$$partial\text{-}covering(z) \equiv (\forall x)(x \in z \rightarrow domino\text{-}on\text{-}board(x))$$
$$\wedge (\forall x \, y)(x \in z \wedge y \in z \rightarrow x = y \vee x \cap y = \{\})$$

Theorem

$$\neg(\exists z)(partial\text{-}covering(z) \wedge \bigcup z = mutilated\text{-}board)$$

Table 5.5 Paulson's 1996 Formalisation

Definitions

$$lessThan(m) := \{i \in \mathbb{N} | i < m\}$$

$$board := lessThan(2 \cdot s(m)) \times lessThan(2 \cdot s(m))$$

$$tiling(A):set(set(\alpha)) := \{\} \in tiling(A) \wedge (a \in A \wedge t \in tiling(a) \wedge a \cap t = \{\})$$
$$\rightarrow a \cup t \in tiling(a)$$

$$domino:set(set(\mathbb{N} \times \mathbb{N})) := \{(i,j),(i,s(j))\} \in domino \wedge$$
$$\{(i,j),(s(i),j)\} \in domino$$

Theorem

$$(board \setminus \{(0,0)\}) \setminus \{(s(2 \cdot m),s(2 \cdot n))\} \notin tiling(domino)$$

Table 5.6 Huet's 1996 Formalisation

$$injective(Board) \wedge injective(Domino) \wedge finite(B) \rightarrow surjective(Domino)$$

Other proofs go further in the generalisation of the theorem. For instance, Paulson generalised the problem to arbitrary $(2n) \times (2n)$ mutilated checkerboards, see Table 5.5. This formalisation has a new quality since no finite check is possible anymore. Some knowledge about set theory and about natural numbers is necessary in order to understand the formalisation and correspondingly the proof.

Also in 1996, Huet formalised the problem, making use of Coq's second order logic. Given a signature of sets B and W, two functions $Board : B \rightarrow W$ and $Domino : W \rightarrow B$ representing the board and the dominoes, the existence of a tiling for the full checkerboard problem is stated as:

Table 5.7 Formalisation in Kerber and Pollet (2006)

Two sets B and W cannot at the same time have the same cardinality and different cardinalities, that is, not $|B| = |W|$ and $|B| \neq |W|$

We can generate from the full checkerboard a mutilated one by taking B' as a proper subset of B. The theorem is then that for an injective function $Board'$ with $Board' : B \rightarrow W$ and $finite(B')$ there is no function $Domino : W \rightarrow B'$ which is injective (i.e. a—possibly partial—covering) and surjective (i.e. total).

Even more abstract is the representation of the problem given presented in Table 5.7, based on the cardinality of sets (finite or infinite).

Essentially, the argument is that if there are strictly fewer black than white elements then we cannot have $|W| = |B|$ (that is, equal numbers which would follow from a covering with 2×1-dominoes).

5.5 Proofs as Communication Between a Writer and a Reader

A very abstract proof—such as the last presented in Sect. 5.3 requires a very intelligent system that can understand it. An even more abstract proof—a proof that can be found very often in mathematics—is: *"This is trivial."* Actually Feynman states only half jokingly that "mathematicians can prove only trivial theorems, because every theorem that's proved is trivial." (Feynman 1985, p.84). Obviously, this is not quite true. Let us consider a proof as information given by the writer of the proof to the reader (and mostly then also checker) of the proof. In order to understand a proof the reader must have some level of understanding and that level of understanding must be at an appropriate level. Immediately after a reader has read and truly understood a proof it is at that moment trivial. The same would hold for a being whose intellect is infinitely powerful as Ayer puts it in (Ayer 1936, p.85f):

> The power of logic and mathematics to surprise us depends, like their usefulness, on the limitations of our reason. A being whose intellect was infinitely powerful would take no interest in logic and mathematics. For he would be able to see at a glance everything that his definitions implied, and, accordingly could never learn anything from logical inference which he was not fully conscious of already. But our intellects are not of this order. It is only a minute proportion of the consequences of our definitions that we are able to detect at a glance. Even so simple a tautology as "91 × 79 = 7189" is beyond the scope of our immediate apprehension. To assure ourselves that "7189" is synonymous with "91 × 79" we have to resort to calculation, which is simply a process of tautological transformation—that is, a process by which we change the form of expression without altering their significance. The multiplication tables are rules for carrying out this process in arithmetic, just as the laws of logic are rules for the tautological transformation of sentences expressed in logical symbolism or in ordinary language.

In the range from *trivial*, over a single hint such as *colour the squares and see whether there are equally many black and white squares*, to a dull enumeration of cases there is a wide range of differing intelligence required. Note that the response to the *"trivial"* proof is not trivial itself. It requires often a lot of intelligence to separate the proofs from the non-proofs and thereby the theorems from the non-theorems. That is, for a reader or checker of a proof, it is not sufficient to always answer "Yes," but they need to also say "No" if the theorem is no theorem (or if the proof is not trivial at all). The detailed dull enumeration of cases assumes that a fixed representation is given and only a few rules need to be applied to check that a proof is a proof indeed (possibly quite a number of rules).

There is more to the concept of proof (see, e.g. Bundy et al. (2005)) than that given by logicians—their assumption that theorem proving can be achieved by a pre-defined set of rules applied to an initial state until a goal state is reached. In order to achieve a human-level understanding of representations and proofs, we need to understand proofs at different levels and also understand their relationships. Human mathematicians learn and extend their repertoire of proof rules (an attempt to get a simple approximation of this can be found in Jamnik et al. (2003)). Humans are good at understanding informal arguments, which by the way can be very precise, and there are attempts to model this with computers (see, e.g. Zinn (2004)).

Ever since I met Sloman first I felt inspired by his view on proofs and the question that goes beyond anything discussed in this contribution: What kind of architecture is necessary in order to detect theorems and come up with a proof (see, e.g. Sloman (1996))?

Obviously human beings go through a certain development and typically reach a level of abstract thinking that allows them to detect theorems and prove them. A proof in this sense is not necessarily a logical proof and not necessarily based on a calculus, and it is also not necessarily 100process that convinces a person (and possibly via communication then also other persons) of the truth of a certain theorem. The development of an architecture that can model this seems still to be a big challenge.

5.6 Conclusion

The view that a proof is just a derivation from axioms applying some proof rules is too simplistic to capture all aspects of a proof. It is a very restricted and reduced model compared to mathematical practice. Human mathematicians have a rich repertoire of proof methods and these cannot be easily captured by a calculus, humans reformulate problems to make them more amenable, they generalise the problem and look for proofs that are simple, since simplicity makes them more beautiful, eases communication and checking, and allows for generalisations. Machines are not yet good at this and it is not clear yet how corresponding systems can be built. In this context, it looks doubtful that a simple architecture will do, but more likely a more complex architecture will be needed. The original idea of AI to address hard problems by trying to model how humans can deal with them may still be the best starting point. An answer to Sloman's question Sloman (1996)—*What sort of architecture is required for a human-like agent?*—may be crucial to making any significant progress in this matter.

References

Ayer AJ (1936) Language, truth and logic, 2nd edn. Victor Gollancz Ltd, London

Bancerek G (1995) The mutilated chessboard problem—checked by MIZAR. In: Matuszewski R (ed) The QED Workshop II, Warsaw University, pp 37–38

Bundy A, Jamnik M, Fugard A (2005) What is a proof? Philos Trans R Soc 363(1835):2377–2391

Descartes R (1637) Discours de la méthode. http://www.gutenberg.org/ebooks/13846

Feynman RP (1985) Surely you're joking Mr. Feynman, Vintage, London

Hardy G (1940) A mathematician's apology. Cambridge University Press, London

Herbrand J (1930) Recherches sur la théorie de la démonstration. Sci Lett Varsovie, Classe III Sci Math Phys 33

Jamnik M, Kerber M, Pollet M, Benzmüller C (2003) Automatic learning of proof methods in proof planning. Logic J IGPL 11(6):647–673

Kerber M, Pollet M (2006) A tough nut for mathematical knowledge management. In: Kohlhase M (ed) Mathematical knowledge management—4th international conference, MKM 2005. Springer, LNAI 3863. Bremen, Germany, pp 81–95

Lakatos I (1976) Proofs and refutations. Cambridge University Press, London

Matuszewski R (ed) (1995) The QED workshop II. http://www.mcs.anl.gov/qed/index.html

McCarthy J (1964) A tough nut for proof procedures. AI Project Memo 16. Stanford University, Stanford, California, USA

McCarthy J (1995) The mutilated checkerboard in set theory. In: Matuszewski (ed) pp 25–26. http://www.mcs.anl.gov/qed/index.html

McCune W (1997) Solution of the Robbins problems. J Autom Reason 19(3):263–276

Robinson JA (1965) A machine oriented logic based on the resolution principle. JACM 12:23–41

Sloman A (1996) What sort of architecture is required for a human-like agent? In: Rao RK (ed) Foundations of rational agency. Kluwer, The Netherlands

Zinn C (2004) Understanding informal mathematical discourse. PhD thesis, Friedrich-Alexander Universität Erlangen-Nürnberg

Chapter 6
What Does It Mean to Have an Architecture?

Brian Logan

6.1 Introduction

The notion of an agent's 'architecture' is a fundamental idea in agent research. Architectures abstract from the details of individual agent programmes, allowing us to make general statements about the properties of a whole class of agents, and hence which kinds of agents are more appropriate for particular environments. From an engineering perspective, choosing an appropriate architecture is therefore a key step in developing an agent to perform a particular task in a specified environment. From a scientific perspective, the notion of an architecture is key to understand the range of possible ways in which (artificial) intelligence can be realised, or how it might have evolved.

An architecture can be thought of as specifying both a set of concepts which can be used to talk about an intelligent system (e.g. a robot or an software agent) and a set of (usually high-level) capabilities realised by that system. For example, if we say that a system has 'beliefs', we are saying something about its internal representations of states of affairs. If we say that it can 'plan', we mean that it is capable of generating a representation of a sequence of future actions, which, if executed in a state of affairs in which its beliefs hold, will achieve its goal(s).

However, despite the importance of architectures in research on agent systems, I will argue the notion of 'an architecture' remains hard to pin down. The agent literature offers little beyond descriptions of particular architectures and some fairly preliminary classifications of architectures and/or architectural components into 'reactive', 'deliberative', 'hybrid' etc. (see, for example, (Nilsson 1998; Sloman and Scheutz 2002)). While such classifications can be useful, it can be difficult to

---

This chapter is a revised and extended version of a paper presented at the 2007 AAAI *Workshop on Evaluating Architectures for Intelligence* (Alechina and Logan 2007).

B. Logan (✉)
School of Computer Science, University of Nottingham, Nottingham NG8 1BB, UK
e-mail: bsl@cs.nott.ac.uk

characterise a particular agent architecture with any precision, or to determine its properties. There are many variants of 'Belief-Desire-Intention' (BDI) architectures in the agent literature, see for example, (Georgeff et al. 1999), however, we lack a framework for their systematic analysis of their properties. This lack of precision makes it hard to compare or evaluate architectures or to say that a given intelligent system has a particular property as a consequence of it having a particular architecture.

In this chapter, I propose an approach to architectures and their analysis which both make precise exactly what it means for a system to 'have' an architecture, and allows us to establish properties of a system expressed in terms of its architectural description. In what follows, I focus on architectures of software agents, however a similar approach can be applied to robots and other kinds of intelligent systems. I show how to establish a precise formal relationship between an architectural property, an agent architecture and an agent model (taken to be an implementation-level description of an agent). A consequence of this approach is that it is possible to verify whether two different agent programmes really have the same architecture or the same properties, allowing a more precise comparison of architectures and the agent programmes which implement them. I illustrate the approach with a number of examples drawn from our previous work at Nottingham, which show how to establish both qualitative and quantitative properties of architectures and agent programmes, and how to establish whether a particular agent has a particular architecture.

6.2 What is an Architecture?

In AI textbooks, the behaviour of an agent is often specified by an *agent function* (also known as an action selection function) which maps a sequence of percepts to an action (see, for example, (Russell and Norvig 2003)). Such an abstractly specified agent function is implemented by an *agent programme*, which may be written in a conventional programming language, or in a special-purpose agent programming language. The agent programme in turn runs on an *agent architecture*. The architecture is a (possibly virtual) machine that makes the percepts from the agent's sensors available to the agent programme, runs the programme and passes the action(s) selected by the programme to the agent's actuators.

In this view, agent programming is conventionally conceived as the problem of synthesising an agent function. For example, Russell and Norvig claim that: 'the job of AI is to design the agent programme that implements the agent function mapping percepts to actions' (Russell and Norvig 2003, p. 44). While this characterisation of AI is correct, it is not very helpful: designing an agent programme that 'implements the agent function mapping percepts to actions' is a very difficult problem. The AI literature describes techniques and algorithms that can be used to solve parts of the problem, however, there is little guidance on how these various components could or should fit together to form an integrated system.

One way of making the problem more tractable is by drawing on the notion of an agent architecture. The notion of 'agent architecture' is ubiquitous in the agent literature but is not well analysed. For example, Russell and Norvig view the architecture as simply 'some sort of computing device with physical sensors and actuators' (Russell and Norvig 2003, p. 44).[1] Moreover, when architectures are discussed, it is often in the context of a particular agent programming language or platform, making it difficult to separate relatively minor language features from the essentials of the architecture.

For the purposes of this chapter, we can think of an architecture as defining a (real or virtual) machine that runs the agent programme. The architecture defines the *atomic operations* of the agent programme and implicitly determines the components of the agent. It determines which operations happen automatically, without the agent programme (or programmer) having to do anything, e.g. the interaction between memory, learning and reasoning. As a result, an architecture constrains kinds of agent programmes we can write (easily). For example, implementing a complex search algorithm in an architecture that supports only simple condition action rules will be more difficult than in an architecture that supports backtracking. In this view, an agent architecture can be seen as defining a class of agent programmes. Just as programmes have properties that make them more or less successful in a given task environment, architectures (classes of programmes) have higher level properties that determine their suitability for a task environment. Choosing an appropriate architecture can therefore make it much easier to develop an agent programme for a particular task environment.

In many cases, it can be difficult to say what 'counts as' the architecture of an agent. A particular agent implementation may consist of a whole hierarchy of virtual machines. For example, 'the' architecture is usually implemented in terms of a programming language, which in turn is implemented using the instruction set of a particular CPU (or a JVM). Likewise some 'agent programmes' together with their underlying architecture can implement a new, higher level architecture (virtual machine), as when, for example, encapsulated robust behaviours in a hybrid architecture effectively define a new set of atomic operations for the deliberative level. In what follows, when used without qualification, 'agent architecture' means the most abstract architecture or the highest-level virtual machine.

How can we make this notion of agent architecture precise? This notion of an 'agent architecture' is clearly related to the more general notion of 'software architecture'. In their classic paper on the foundations of software architectures, Perry and Wolf characterise software architecture as: 'concerned with the selection of architectural elements, their interactions, and the constraints on those elements and their interactions necessary to provide a framework in which to satisfy the requirements and serve as a basis for the design' (Perry and Wolf 1992, p. 43). A similar view is advanced in a recent survey by Shaw and Clements, who define software architectures as the 'principled study of the large-scale structures of software systems' (Shaw and Clements 2006, p. 31). However, while there has been considerable work in software engineering resulting in standard architectures for many domains and

---

[1] Although Russell and Norvig (2003) state that 'agent = architecture + programme', most of the book is about designing (components of) agent programmes, and there are only ten references to 'architecture' in nearly 1,000 pages.

applications, e.g. *n*-tier client-server architectures, service-oriented architectures, etc., I would argue (claims in (Shaw and Clements 2006) notwithstanding) that the notion of architecture as used in software engineering is hardly any more precise than that used in AI.

 The notion of an agent architecture is also related to the notion of a 'cognitive architecture' as used in artificial intelligence and cognitive science. A cognitive architecture can be defined as an integrated system capable of supporting intelligence. The term is often used to denote models of human reasoning, e.g. ACT-R (Anderson and Libiere 1998), SOAR (Newell 1990). However, in other cases no claims about psychological plausibility are made. In this latter sense, 'cognitive architecture' is more or less synonymous with 'agent architecture', as used here. However, while a number of cognitive architectures have been extensively studied, there has been relatively little work on understanding how different cognitive architectures relate to each other, or on which features of an architecture are responsible for its performance on a particular task.

 I believe that what is required is a more precise characterisation of what an 'architecture' is and what it means for an agent to have one. In the remainder of this chapter, I sketch out one approach to a more formal characterisation of agent architectures that draws heavily on joint work with colleagues and students at Nottingham and elsewhere.

6.3 Specifying Architectures and Properties

An architecture is a way of looking at or conceptualising an agent programme that defines the possible states of the agent and transitions between them. We can think of an architecture as providing a language of concepts and relations for describing the contents of agent states and their possible transitions. Different architectures characterise the contents of the agent's state in different ways, e.g. as beliefs or goals or 'affective' states such as 'hungry' or 'fearful', or simply in terms of the contents of particular memory locations. Similarly, the possible transitions may be characterised in terms of basic agent capabilities, such as performing a 'basic action', or by higher level capabilities such as 'perception', 'learning' or 'planning'. Each architecture exposes some of the contents of the states and some of the possible transitions to the agent developer. For example, an agent programming language or toolkit may allow the representation of beliefs about any state of affairs, whereas a particular agent implementation may only have beliefs about state occurring in a particular task environment or problem domain. Those transitions which are not under the control of the developer typically correspond to the execution of some 'basic' control cycle specified by the architecture. For example, a particular architecture may specify that the execution of a basic action or plan step is always followed by a 'sensing' step.

 In this view, the 'pure' architecture of an agent consists of a language for describing its states and transitions, and a set of constraints which limit the possible states and the transitions between them to those allowed by the architecture. For example, a possible constraint on the beliefs of an agent imposed by its architecture may be that its beliefs are always consistent, or that they always include, e.g. a belief about the current time or the agent's battery level.

Since instances of an architecture are computer programmes, they can be represented as state transition systems. The states of the state transition system will be determined by the possible states defined by the architecture. Similarly, the transitions between states are determined by the kinds of transitions supported by the architecture. The set of all possible transition systems corresponding to instances of an architecture provides a formal model of the architecture itself. Note that we assume that this set is not given extensionally, but is defined by a set of constraints on possible states and possible transitions between states.

The key idea underlying this approach is to define a logic which axiomatises the set of transition systems corresponding to the architecture of interest. This logic can then be used to state properties of the architecture (typically couched in terms of the concepts defined by the architecture). An architecture has a particular property if a formula in the logic expressing the property is true in all transition systems defined by the architecture. For example, we can formulate precisely in a suitable logic (e.g. CTL or PDL extended with belief and goal modalities) various *qualitative* properties of architectures such as, 'Can an agent have inconsistent beliefs?', 'Can an agent drop a goal?' or 'Can an agent drop a goal which is not achieved?', and check which of these properties are satisfied by which architectures. Perhaps more surprisingly, we can also use this approach to evaluate various *quantitative* properties of architectures. For example, 'how much time would choosing an action take in this agent architecture?', or 'how much memory is required to perform deliberation?' Note that by appropriate choice of the logic, we can perform an evaluation at different levels of detail. For example we can establish properties of the 'pure' architecture, or of a particular agent programme, or a particular agent programme with particular inputs.

There exists a substantial amount of work on modelling agent systems using state transition systems and expressing correctness properties in logic (see, for example, (Rao and Georgeff 1991; Fagin et al. 1995; van der Hoek et al. 1999; Wooldridge 2000)). The major difference between this work and the approach advocated here, is our use of syntactic belief, goal etc. modalities rather than a standard Kripke-style semantics to model agent beliefs and other propositional attitudes, thus avoiding the problem of logical omniscience (that the agents are modelled as believing all logical consequences of their beliefs) (Hintikka 1962). Since reasoning takes time and memory in any implemented agent system, formal models based on possible worlds semantics may incorrectly ascribe properties to real agent systems.

In the remainder of this chapter I illustrate this approach by means of some simple examples drawn from our previous work.

6.4 Correctness Properties

In this section I briefly survey some of our previous works in establishing the correctness of agent programmes and show how the same approach can be lifted to the level of agent architectures. In (Alechina et al. 2007, 2010b) we proposed a sound and complete logic for agent programmes written in SimpleAPL, an APL-like (Dastani et al. 2004; Dastani and Meyer 2008) agent programming language. The logic is a variant of Propositional Dynamic Logic (PDL) (Fischer and Ladner 1979) extended with belief and goal operators, which allows the specification of safety and liveness properties of SimpleAPL agent programmes. In that work, we showed how to define transition systems corresponding to different programme execution strategies in logic, however, the execution strategies were 'hard-coded' into the translation of agent programmes. In (Alechina et al. 2008a, 2010a) we extended this approach to define transition systems corresponding to different programme execution strategies using a single-fixed translation of the agent programme together with an axiomatisation of the agent's execution strategy. This more abstract formalisation makes it much easier to analyse the implications of a key aspect of the agent's architecture — its deliberation strategy. As an illustration, we showed how the choice of deliberation strategy can determine whether an agent is able to achieve it goal(s). In more recent work (Doan et al. 2009; Alechina et al. 2011), we have extended this approach further to consider agents which are able to modify their plans at run time. We consider the plan revision rules of the agent programming languages Dribble (van Riemsdijk et al. 2003), 3APL (Dastani et al. 2004, 2005) and 2APL (Dastani and Meyer 2008; Dastani 2008) which allow an agent to revise its plans at run time in response to changes in its environment and the effects nondeterministic action execution. We show how such agents can be formalised in logic, and how to verify safety properties of agent programmes that allow plan revision. This work can be seen as a (very preliminary) formalisation of a basic reflective capability, i.e. the ability of an agent to modify a chosen course of action in response to unanticipated opportunities or failures in the execution of a plan.

6.5 Architectural Properties

While our previous work has touched on architectural issues such an agent's deliberation strategy and simple reflective capabilities, our focus was mainly on the correctness of agent *programmes* implemented using these architectures. In this section, I will briefly outline how the same approach can be applied to 'pure' agent *architectures*, using the SimpleAPL architecture as an example. I first briefly describe the language and architecture of SimpleAPL and then sketch how to formalise them in logic.

SimpleAPL is a fragment of 3APL (Dastani et al. 2004, 2005). A SimpleAPL agent has a set of goals (ground literals), beliefs (ground literals), plans and planning goal rules. A planning goal rule of the form

$$\psi \leftarrow \phi \mid \pi$$

can be read as 'if you have a goal ψ and you believe ϕ, then adopt plan π'. SimpleAPL plans are pre-defined by the programmer, and are built from basic actions using sequential composition, conditionals and iteration. Basic actions have a finite set of pre- and post-conditions. For simplicity, to avoid modelling the environment, we assume that the agent's beliefs are always correct and actions always successful, which allows us to express pre- and post-conditions in terms of the agent's beliefs. The SimpleAPL agent architecture attempts to achieve the agent's goals by selecting appropriate plans given the agent's beliefs. A SimpleAPL agent may have several active plans at any given time. At each execution cycle, the agent can either apply a planning goal rule to select a new plan, or execute the first step in any of its current plans.

6.5.1 Operational Semantics

We can define the formal semantics of SimpleAPL in terms of a transition system. Each transition corresponds to a single execution step and takes the agent from one configuration to another. Configurations consist of the beliefs, goals and plans of the agent.

Definition 1 The configuration of an agent is defined as $\langle \sigma, \gamma, \Pi \rangle$ where σ is a set of literals representing the agent's beliefs, γ is a set of literals representing the agent's goals, and Π is a set of plan entries representing the agent's current active plans.

An agent's initial beliefs and goals are specified by its programme, and Π is initially empty. Executing the agent's programme modifies its initial configuration in accordance with a set of transition rules. Below, we give only the transition rule for basic actions; for the full operational semantics see (Alechina et al. 2007).

6.5.1.1 Basic Actions

A basic action α can be executed if its pre-condition is entailed by the agent's beliefs, ie. $\sigma \models \phi$. Executing the action adds the literals in the postcondition to the agent's beliefs and removes any existing beliefs which are inconsistent with the postcondition.

$$\frac{\alpha ; \pi \in \Pi \quad T(\alpha, \sigma) = \sigma' \quad \gamma' = \gamma \setminus \{\phi \in \gamma \mid \sigma' \models \phi\}}{\langle \sigma, \gamma, \Pi \rangle \longrightarrow \langle \sigma', \gamma', \Pi \setminus \{\alpha ; \pi\} \cup \{\pi\} \rangle}$$

T is a partial function that takes a belief update action α and a belief set σ, and returns the modified belief set if the precondition of the action is entailed by σ. Note that executing a belief update action causes the agent to drop any goals it believes to be achieved as a result of the update.

6.5.2 Logic

We can define a logic which allows us to specify properties of the SimpleAPL agent architecture. We begin by defining transition systems which capture the *capabilities*

of agents as specified by their basic actions.[2] We then show how to interpret a variant
of the temporal logic CTL with belief and goal operators in this semantics.

6.5.2.1 States and Transitions

Let P denote the set of propositional variables used to describe agent's beliefs and
goals. A state s is a pair $\langle \sigma, \gamma \rangle$, where:

σ is a set of beliefs $\{(-)p_1, \ldots, (-)p_n : p_i \in P\}$. We assume that belief states are
consistent, i.e. for no $p \in P$ both p and $-p \in \sigma$.

γ is a set of goals $\{(-)u_1, \ldots, (-)u_n : u_i \in P\}$. The set of goals does not have
to be consistent, but it has to be disjoint from σ: no element of γ is in σ (i.e. is
already believed).

Let the set of basic actions be $\text{Ac} = \{\alpha_1, \ldots, \alpha_m\}$. We associate with each $\alpha_i \in$
Ac a set of pre- and postconditions of the form $\{(-)p_1 \in \sigma, \ldots, (-)p_n \in \sigma\}$,
$\{(-)q_1 \in \sigma', \ldots, (-)q_k \in \sigma'\}$ (where ps and qs are not necessarily disjoint) which
mean: if α_i is executed in a state with belief set σ that satisfies the precondition, then
the resulting state s' has the belief set σ' that satisfies the postcondition (including
replacing p with $-p$ if necessary to restore consistency), *the rest of σ' is the same
as σ*, and the goal set $\gamma' = \gamma \setminus \{(-)p : (-)p \in \sigma'\}$.

Executing an action α_i in different configurations may give different results. For
each α_i, we denote the set of pre- and postcondition pairs $\{(\text{prec}_1, \text{post}_1), \ldots,$
$(\text{prec}_l, \text{post}_l)\}$ by $C(\alpha_i)$. We assume that $C(\alpha_i)$ is finite, that pre-conditions
$\text{prec}_j, \text{prec}_k$ are mutually exclusive if $j \neq k$, and that each precondition has
exactly one associated postcondition. We denote the set of all pre- and postconditions
by $\mathbf{C}$.

6.5.2.2 Language

The language L for talking about the agent's beliefs, goals and plans is the language
of CTL extended with belief operator B and goal operator G. A formula of L is
defined as follows: if $p \in P$, then $B(-)p$ and $G(-)p$ are formulas; if ϕ and ψ
formulas, then $EX\phi$, $E\Box\phi$,[3] and $EU(\phi, \psi)$ are formulas; and L is closed under the
usual boolean connectives.

---

[2] Note that the transition systems are more general than operational semantics of the SimpleAPL
architecture presented informally above, in that they do not describe a particular agent programme
or execution strategy, but all possible basic transitions between all the beliefs and goal states of an
agent.

[3] Note that we use $\Box$ rather than the more conventional G to avoid confusion with the goal operator.

6.5.2.3 Semantics

A model for L is a structure $M = (S, \{R_{\alpha_i} : \alpha_i \in \text{Ac}\}, V)$, where

- S is a set of states.
- $V = (V_b, V_g)$ is the evaluation function consisting of belief and goal valuation functions V_b and V_g such that for $s = \langle \sigma, \gamma \rangle$, $V_b(s) = \sigma$ and $V_g(s) = \gamma$.
- R_{α_i}, for each $\alpha_i \in \text{Ac}$, is a relation on S such that $(s, s') \in R_{\alpha_i}$ iff for some $(\text{prec}_j, \text{post}_j) \in C(\alpha_i)$, $\text{prec}_j(s)$ and $\text{post}_j(s')$, i.e. for some pair of pre- and postconditions of α_i, the precondition holds for s and the corresponding post-condition holds for s'. Note that this implies two things: first, an α_i transition can only originate in a state s which satisfies one of the preconditions for α_i; second, since pre-conditions are mutually exclusive, every such s satisfies exactly one pre-condition, and all α_i-successors of s satisfy the matching post-condition.

The relation $\models$ of a formula being true in a state of a model is defined inductively as follows:

- $M, s \models B(-)p$ iff $(-)p \in V_b(s)$
- $M, s \models G(-)p$ iff $(-)p \in V_g(s)$
- $M, s \models \neg\phi$ iff $M, s \not\models \phi$
- $M, s \models \phi \wedge \psi$ iff $M, s \models \phi$ and $M, s \models \psi$
- $M, s \models EX\phi$ iff on some computation path starting in s ϕ holds in the next state
- $M, s \models E\Box\phi$ iff there exists a computation path starting in s such that ϕ holds globally (in every state along the path including s)
- $M, s \models EU(\phi, \psi)$ iff there exists a computation path starting in s such that ϕ holds in every state along the path (including s) until ψ holds

Let the class of transition systems defined above be denoted $\mathbf{M_P}$ (note that $\mathbf{M}$ is parameterised by the set $\mathbf{P}$ of all possible SimpleAPL programmes).

In this logic, we can state properties of the SimpleAPL architecture and prove (by semantic argument) that they are valid (true in all models). For example, we can show that the same formula cannot be both a goal and a belief

$$A\Box\neg(B\phi \wedge G\phi)$$

where $A\Box\psi \equiv \neg EU(\top, \neg\psi)$. More interestingly, we can show that a SimpleAPL agent is blindly or fanatically committed to to its goals (Rao and Georgeff 1992), i.e. it will only abandon a goal if it believes it has been achieved

$$G\phi \rightarrow A\Box(G\phi \vee B\phi)$$

These examples are very simple. However, they serve to illustrate how this approach allows the precise specification of properties of an agent architecture.

6.6 Resource Requirements

We can use a similar approach to compare the relative resource requirements of architectures. For example, given two agents with different architectures, how much time or memory do they require to solve the same problem? Or, given two multi-agent systems possessing the same information, how many messages do the agents in each system have to exchange before they solve a given problem? We have studied these problems for rule-based agents (Alechina et al. 2004a, b; Alechina and Logan 2005; Alechina et al. 2006b) and for agents reasoning in a variety of logics (Alechina et al. 2006a; Albore et al. 2006; Alechina et al. 2008b).

An important aspect of the architecture of a rule-based agent is how the rule instances which are applied to produce the next state are selected from the set of all matching rule instances (i.e. the agent's conflict resolution strategy). This can affect the time (number of rule application cycles) that an agent takes to select an action or to assert a certain fact in memory. In (Alechina et al. 2004a, b) we axiomatised classes of transition systems corresponding to different kinds of conflict resolution strategies in Timed Reasoning Logic (TRL). TRL attaches time labels to formulas, allowing us to express in the logic the fact that a certain statement will be derived by the agent within n timesteps. For example, in (Alechina et al. 2004b), we used TRL to model two rule-based agents with the same set of rules and the same set of initial observations, but with two different conflict resolution strategies. One agent used the *depth* conflict resolution strategy of CLIPS, which favours more recently derived information, while the other used a *breadth* conflict resolution strategy which favours older information. We showed how the times at which facts are derived differ for the two agents, and also that facts will be derived sooner if the two agents communicate.

As well as considering time requirements of agents with different architectures, we can also consider their respective memory requirements. A simple measure of the agent's memory requirement is the size of the agent's belief base, which can be identified either with the number of distinct beliefs in the belief base, or the total number of symbols required to represent all beliefs in the belief base. Logics for agents with bounded memory were studied in (Alechina et al. 2006a; Albore et al. 2006), where we showed the existence of trade-offs between the agent's time and memory requirements: for example, deriving the same conclusion with a smaller memory may require more derivation steps. In (Alechina et al. 2008b) we considered the interaction of time, memory and communication in systems of distributed reasoning agents, and showed that the agents can achieve a goal only if they are prepared to commit certain time, memory and communication resources.

6.7 Correct Implementation

The approach outlined above allows us to study properties of architectures in the abstract and of agent programmes expressed in terms of architecture-specific concepts. However, it leaves open the question of whether a particular agent programme described in implementation-level terms can be said to realise a particular architecture. In this section we sketch what it means in our view for an agent to implement or realise an architecture, and give examples in which we show how to correctly ascribe beliefs to a behaviour-based agent, and how to model the ascription of beliefs in systems of agents which reason about each others' beliefs.

Given our definition of (a formal model of) an architecture as the set of all transition systems corresponding to programmes which implement the architecture, there is a trivial sense in which a particular agent programme can be seen as implementing an architecture: this holds if a transition system corresponding to the programme is in the set of transition systems corresponding to the architecture. However, in practice the problem is often more complex, in that the 'natural' descriptions of the architecture and its implementation may use different concepts and levels of detail. For example, the language for specifying constraints on transition systems corresponding to a BDI architecture would involve ways of referring to beliefs, desires and intentions, while the description of an agent programme may be in terms of basic programming constructs and data structures. We therefore define the notion of 'implementation' in terms of a mapping between the transition systems corresponding to an agent programme and an architecture, and say that a particular agent implements a particular architecture if the transition system describing the agent programme can be mapped into one of the transition systems in the architecture set. For example, if the low-level model of an agent involves two boolean state variables x and y, we may decide to translate $x = 1$ as 'the agent has belief p' and $y = 1$ as 'the agent has goal q'. Similarly, we may collapse several of the agent's low-level actions into a single 'basic action' at the architecture level. If such a mapping into one of the transition systems corresponding to an architecture can be carried out (namely, there exists a transition system S in the set corresponding to the architecture, such that initial states of the agent translate into initial states of S, and every time when there is a sequence of low-level transitions in the agent's description, then there is a corresponding basic action in S, again leading to the matching states), we will say that the agent programme implements the architecture.

This 'implementation relation' defines precisely when a given agent can be said to have or implement a particular architecture and hence satisfies all the properties of the architecture. Clearly, the same agent programme may implement several different architectures under this definition. Indeed an agent can correctly be said to implement any architecture for which there exists a mapping from the agent model to the architectural description.

As an example, in (Alechina and Logan 2002) we showed how to correctly ascribe beliefs to agents which do not employ explicit representations of beliefs, e.g. where the behaviour of the agent is controlled by a collection of decision rules or reactive behaviours which simply respond to sensor inputs from the agent's environment. In particular, we showed that if we only ascribe an agent beliefs in literals then it is safe to model the agent's beliefs in a logic such as KD45 or S5 in which beliefs are modelled as closed under logical consequence. (KD45 and S5 are widely used in formal models of agents, see, e.g. (Rao and Georgeff 1991; Fagin et al. 1995; van der Hoek et al. 1999; Wooldridge 2000). However, it is not safe to ascribe beliefs in complex formulas to an agent, because the actual state of the agent (which does not contain all the consequences of its beliefs) and the corresponding formal model (which assumes its beliefs to be logically closed) would not match. In (Alechina and Logan 2009) we extended this approach to model the 'sense', 'think' and 'act' phases of an agent's deliberation cycle, allowing correct ascription of beliefs at each stage of the agent's execution. In (Alechina and Logan 2010) we considered belief ascription for systems of agents which reason about each other's beliefs. We showed that for agents whose computational resources and memory are bounded, correct ascription of beliefs cannot be guaranteed, even in the limit. We proposed a solution to the problem of correct belief ascription for feasible agents which involves ascribing reasoning strategies, or preferences on formulas, to other agents and showed that if a resource-bounded agent knows the reasoning strategy of another agent, then its ascription of beliefs to the other agent is correct in the limit.

This work could be seen as extension (Sloman 1994) in explicating the ways in which 'architecture dominates mechanism'. However, I believe it can also help us to understand the ways in which a single mechanism may simultaneously realise multiple architectures.

6.8 Summary

I have presented a methodology for the formal characterisation and analysis of agent architectures. The methodology consists of defining a set of transition systems corresponding to an architecture of interest, and verifying properties of this set of transition systems. A key feature of this approach is the use of syntactic belief, goal, etc. modalities rather than a standard Kripke-style semantics to model the beliefs and other propositional attitudes of agents, so avoiding the problem of logical omniscience. The characterisation is therefore of the architecture in a very direct sense. The results of the analysis hold for a particular agent (relative to some implementation-level description of the agent) insofar as the agent implements the architecture. While such an analysis does not allow us to answer all questions of interest regarding architectures, I believe it allows us to address a (surprisingly) wide range of architectural issues. As examples, I sketched to how establish correctness properties, quantitative properties and how to correctly ascribe beliefs to agents. However, I believe this is a potentially very productive area, and our previous work has barely begun to scratch the surface of what it means to have an architecture.

Acknowledgments I am highly indebted to Natasha Alechina: without her logical abilities and willingness to invent nonstandard logical approaches to models of agents, none of this work would have been possible. I am also indebted to colleagues in the agent programming language community, particularly Rafael Bordini, Mehdi Dastani, Koen Hindriks and John-Jules Meyer and to former and current students in the Agents Lab at Nottingham, including Mark Jago, Neil Madden, Abdur Rakib, Nguyen Hoang Nga, Fahad Khan and Doan Thu Trang.

References

Albore A, Alechina N, Bertoli P, Ghidini C, Logan B, Serafini L (2006) Model-checking memory requirements of resource-bounded reasoners. In: Proceedings of the twenty-first national conference on artificial intelligence (AAAI 2006), AAAI Press, pp 213–218

Alechina N, Logan B (2002) Ascribing beliefs to resource bounded agents. In: Proceedings of the first international joint conference on autonomous agents and multi-agent systems (AAMAS 2002), vol 2. ACM Press, Bologna, pp 881–888

Alechina N, Logan B (2005) Verifying bounds on deliberation time in multi-agent systems. In: Gleizes MP, Kaminka G, Nowe A, Ossowski S, Tuyls K, Verbeeck K (eds) Proceedings of the third European workshop on multiagent systems (EUMAS'05). Koninklijke Vlaamse Academie van Belgie voor Wetenschappen en Kunsten, Brussels, Belgium, pp 25–34

Alechina N, Logan B (2007) Formal evaluation of agent architectures. In: Kaminka GA, Burghart CR (eds) Evaluating architectures for intelligence: papers from the 2007 AAAI workshop, AAAI Press, technical report WS-07-04, pp 1–4

Alechina N, Logan B (2009) A logic of situated resource-bounded agents. J Logic Lang Inf 18(1): 79–95. http://dx.doi.org/10.1007/s10849-008-9073-6

Alechina N, Logan B (2010) Belief ascription under bounded resources. Synthese 173(2):179–197. http://dx.doi.org/10.1007/s11229-009-9706-6

Alechina N, Logan B, Whitsey M (2004a) A complete and decidable logic for resource-bounded agents. In: Jennings NR, Sierra C, Sonenberg L, Tambe M (eds) Proceedings of the third international joint conference on autonomous agents and multi-agent systems (AAMAS 2004), vol 2. ACM Press, New York, pp 606–613

Alechina N, Logan B, Whitsey M (2004b) Modelling communicating agents in timed reasoning logics. In: Alferes JJ, Leite J (eds) Proceedings of the ninth European conference on logics in artificial intelligence (JELIA 2004), Springer, Lisbon, no. 3229 in LNAI, pp 95–107

Alechina N, Bertoli P, Ghidini C, Jago M, Logan B, Serafini L (2006a) Verifying space and time requirements for resource-bounded agents. In: Edelkamp S, Lomuscio A (eds) Proceedings of the fourth workshop on model checking and artificial intelligence (MoChArt-2006), pp 16–30

Alechina N, Jago M, Logan B (2006b) Modal logics for communicating rule-based agents. In: Brewka G, Coradeschi S, Perini A, Traverso P (eds) Proceedings of the 17th European conference on artificial intelligence (ECAI 2006), IOS Press, pp 322–326

Alechina N, Dastani M, Logan B, Meyer JJC (2007) A logic of agent programs. In: Proceedings of the twenty-second AAAI conference on artificial intelligence (AAAI 2007), AAAI Press, pp 795–800

Alechina N, Dastani M, Logan B, Meyer JJC (2008a) Reasoning about agent deliberation. In: Brewka G, Lang J (eds) Proceedings of the eleventh international conference on principles of knowledge representation and reasoning (KR'08). AAAI, Sydney, Australia, pp 16–26

Alechina N, Logan B, Nga NH, Rakib A (2008b) Verifying time, memory and communication bounds in systems of reasoning agents. In: Padgham L, Parkes D, Müller J, Parsons S (eds) Proceedings of the seventh international conference on autonomous agents and multiagent systems (AAMAS 2008), vol 2. IFAAMAS, Estoril, Portugal, pp 736–743

Alechina N, Dastani M, Logan B, Meyer JJC (2010a) Reasoning about agent deliberation. Auton Agent Multi-Agent Syst 22(2):1–26. doi:10.1007/s10458-010-9129-2, http://dx.doi.org/10.1007/s10458-010-9129-2

Alechina N, Dastani M, Logan B, Meyer JJC (2010b) Using theorem proving to verify properties of agent programs. In: Dastani M, Hindriks KV, Meyer JJC (eds) Specification and verification of multi-agent systems, chap 2. Springer, Berlin, pp 1–34

Alechina N, Dastani M, Logan B, Meyer JJC (2011) Reasoning about plan revision in BDI agent programs. Theoretical Computer Science. doi:10.1016/j.tcs.2011.05.052, (in press)

Anderson JR, Libiere C (1998) The atomic components of thought. Lawrence Erlbaum Associates, Lawrence

Dastani M (2008) 2APL: a practical agent programming language. Auton Agent Multi-Agent Syst 16(3):214–248

Dastani M, Meyer JJC (2008) A practical agent programming language. In: Dastani M, El Fallah-Seghrouchni A, Ricci A, Winikoff M (eds) Proceedings of the fifth international workshop on programming multi-agent systems (ProMAS'07),vol 4908. Springer, LNCS, Berlin, pp 107–123

Dastani M, van Riemsdijk MB, Dignum F, Meyer JJC (2004) A programming language for cognitive agents: goal directed 3APL. In: Dastani M, Dix J, El Fallah-Seghrouchni A (eds) Programming multi-agent systems, first international workshop, ProMAS 2003, Melbourne, Australia, July 15, 2003, Selected revised and invited papers, vol 3067. Springer, LNCS, pp 111–130. http://www.springerlink.com/content/l7dqkvqh5u94114b

Dastani M, van Riemsdijk MB, Meyer JJC (2005) Programming multi-agent systems in 3APL. In: Bordini RH, Dastani M, Dix J, El Fallah-Seghrouchni A (eds) Multi-agent programming: languages, platforms and applications, multiagent systems, Artificial Societies, and Simulated Organizations, vol 15. Springer, pp 39–67

Doan TT, Logan B, Alechina N (2009) Verifying dribble agents. In: Baldoni M, Bentahar J, Lloyd J, van Riemsdijk MB (eds) Seventh international workshop on declarative agent languages and technologies (DALT 2009). Workshop Notes, Budapest Hungary, pp 162–177

Fagin R, Halpern JY, Moses Y, Vardi MY (1995) Reasoning about knowledge. MIT Press, Cambridge

Fischer MJ, Ladner RE (1979) Propositional dynamic logic of regular programs. J Comput Syst Sci 18(2):194–211

Georgeff MP, Pell B, Pollack ME, Tambe M, Wooldridge M (1999) The Belief-Desire-Intention model of agency. In: Müller JP, Singh MP, Rao AS (eds) Proceedings Intelligent agents V, agent theories, architectures, and languages, 5th international workshop, (ATAL'98), Paris, France, July 4–7, 1998, vol 1555. Springer, Lecture Notes in Computer Science, pp 1–10

Hintikka J (1962) Knowledge and belief. Cornell University Press, Ithaca

van der Hoek W, van Linder B, Meyer JJC (1999) An integrated modal approach to rational agents. In: Wooldridge M, Rao A (eds) Foundations of rational agency. Kluwer Academic, Dordrecht, pp 133–168

Newell A (1990) Unified theories of cognition. Harvard University Press, Cambridge

Nilsson N (1998) Artificial intelligence: a new synthesis. Morgan Kaufmann Publishers, San Francisco

Perry DE, Wolf AL (1992) Foundations for the study of software architecture. SIGSOFT Softw Eng Notes 17:40–52. http://doi.acm.org/10.1145/141874.141884

Rao AS, Georgeff MP (1991) Modeling rational agents within a BDI-architecture. In: Proceedings of the second international conference on principles of knowledge representation and reasoning (KR'91), pp 473–484

Rao AS, Georgeff MP (1992) An abstract architecture for rational agents. In: Rich C, Swartout W, Nebel B (eds) Proceedings of knowledge representation and reasoning (KR&R-92), pp 439–449

van Riemsdijk B, van der Hoek W, Meyer JJC (2003) Agent programming in Dribble: from beliefs to goals using plans. In: Proceedings of the second international joint conference on autonomous agents and multiagent systems (AAMAS'03), ACM Press, New York, NY, USA, pp 393–400. http://doi.acm.org/10.1145/860575.860639

Russell S, Norvig P (2003) Artificial intelligence: a modern approach, 2nd edn. Prentice Hall, Upper Saddle River

Shaw M, Clements P (2006) The golden age of software architecture. IEEE Softw 23:31–39. doi: 10.1109/MS.2006.58

Sloman A (1994) Semantics in an intelligent control system. Philos Trans R Soc Phys Sci Eng 349(1689):43–58

Sloman A, Scheutz M (2002) A framework for comparing agent architectures. In: Proceedings of the UK workshop on computational intelligence UKCI'02, pp 169–176

Wooldridge M (2000) Reasoning about rational agents. MIT Press, Cambridge

Chapter 7
Virtual Machines: Nonreductionist Bridges Between the Functional and the Physical

Matthias Scheutz

7.1 Introduction

The notion of supervenience has become a major explanatory device for tackling the mind–body problem, in particular, and for relating higher level properties to lower level properties, in general. As (Kim 1998, p. 9) puts it: "Supervenience is standardly taken as a relation between two sets of properties, the supervenient properties and their base properties" (p. 9). Especially, for those who espouse nonreductive physicalism, supervenience seems to provide a solution to *the mind–body problem*: by claiming that mental properties *supervene* on physical properties, the dependence of mental properties on physical properties is secured, and the possibility of changes in the mental without changes in the physical are prevented. At the same time, this dependence does not necessarily lead to the reduction of mental properties to physical properties, as supervenience (at least in some of its readings) is compatible with substance dualism as well as token physicalism.

However, for "property supervenience" to be a potential candidate for a "solution" to the mind–body problem, two crucial assumptions about mental properties need to made: first, that mental properties are properties of physical systems (which means that "being in pain," for example, is a property of a particular patch of four-dimensional spacetime occupied by a creature capable of having pain). And secondly, that the mind–body problem reduces to a problem about a relation between mental and physical properties. Yet, it is not obvious that either assumption is justified, since *prima facie* mind and brain seem to be more than a mere collection of properties, given that psychological entities such as thoughts, imaginations, experiences, etc., are of a very different nature from cells, chemicals, potentials, etc. Thus, to say that the quantifiers in a psychological theory range over the same entities as the quantifiers in a physical theory (which has to be effectively assumed by proponents of

M. Scheutz (✉)
Department of Computer Science, Tufts University, Medford, MA 02155, USA
e-mail: matthias.scheutz@tufts.edu

J. L. Wyatt et al. (eds.), *From Animals to Robots and Back: Reflections on Hard Problems in the Study of Cognition*, Cognitive Systems Monographs 22, DOI: 10.1007/978-3-319-06614-1_7, © Springer International Publishing Switzerland 2014

property supervenience) seems to be a claim that at the very least needs to be argued for.

We believe that such an argument might turn out to be very difficult, if not impossible, and that instead of restricting the notion of supervenience to properties, it might be more promising to talk about *whole ontologies* supervening on other ontologies (see Sloman 1998). And while we certainly do not know yet how this link is affected in the case of mind and brain, we can find paradigmatic examples of different causally interacting ontological levels in the computer science notion of a *virtual machine*, i.e., the formal specifications of computational architectures using a particular ontology. Using the notion of virtual machines being implemented in other virtual or physical machines, we will argue that the notion of *implementation* (of a virtual machine) might elucidate nonreductive dependence in at least one way that property supervenience cannot accomplish: A being implemented in B does not only establish a dependence of A on B, but it also shows *how*, in addition to *that*, A depends on B. Hence, we suggest that the notion of implementation might turn out to be able to do more work in attempts to solve the mind–body problem or how higher level ontologies are realized in lower level ontologies than any of the current notions of (property) supervenience, for notions of supervenience can ultimately only state the mind problem, while the notion of implementation might be able to solve it.

7.2 Supervenience and the Mind–Body Problem

The mind–body problem posed as the question "how does the mind relate to the body?" is very general in nature. It does not per se imply anything about the underlying ontologies, neither the mental nor the bodily. One common answer to the above question is that the mental is determined by and dependent on the physical. This relationship is expressed in term of supervenience, saying that the mental supervenes on the physical. Usually, supervenience is defined as a relation between properties, but that shifts the original problem to "how do mental properties relate to physical properties." The so-obtained restriction, however, seems unnecessary, if not unwarranted, unless it can be shown that the relation between mental and physical ontologies somehow reduces to a relation between certain classes of properties. This, however, does not seem to be the case. First of all, relations cannot be completely reduced to properties (this is a well-known fact from logic). Furthermore, even if it is possible to view some mental relations in terms of a particular set of physical properties, the language will be unnecessarily complicated by avoiding to talk about physical relations. Finally, the general question remains why mental entities such as beliefs, plans, hunches, discoveries, etc., need to be treated as "properties." We believe that this restriction is an alleged consequence of the physicalist ontological substance monism, which does not allow for any nonphysical entities. Alleged, because it does not apriori follow that entities have to be automatically treated as properties of physical spacetime simply because they are physically realized/realizable. In fact, it would not make sense at all to treat numbers as properties of physical objects. As Cantor has reminded us, numbers, *qua* entities, are properties of sets, i.e., nonphysical objects and as such they are not properties of any physical spacetime entity. Yet, one does not usually dispense with numbers or sets for that matter simply because they are abstract entities (unless one is a committed nominalist, but then there are other difficulties). So, there is at least no apriori reason why mental entities have to be viewed as in some sense "reduced" to properties that supervene on physical properties and can be instantiated by them. We believe that keeping mental entities as such in our ontology does not do any harm to the physicalist program of viewing material objects with mentality as being entirely composed of "physical stuff."

7.2.1 Types of Supervenience

Several types of supervenience have been distinguished in the literature, the most common ones being "weak supervenience," "strong supervenience," and "global supervenience."[1] Whereas weak and strong supervenience apply to sets of properties, global supervenience applies to properties of whole worlds. Fixing two sets of properties α and β, then these notions can be defined as follows: α-properties supervene on β-properties if and only if

(Weak supervenience) necessarily, for any mental property $A \in \alpha$, if anything has A, there exists a property $B \in \beta$ it has B, and anything that has B has A.
(Strong supervenience) necessarily, for any mental property $A \in \alpha$, if anything has A, there exists a property $B \in \beta$ it has B, and necessarily anything that has B has A.[2]
(Global supervenience) for any worlds w_i and w_j, if w_i and w_j are β-indiscernible, then w_i and w_j are α-indiscernible.

While the relation between strong and weak supervenience is clearly that of implication (i.e., strong implies weak), the relation of both kinds to global supervenience is not all that clear. Some have argued that global and strong supervenience are equivalent (e.g., Kim 1984), others claim that global supervenience is even weaker than weak supervenience (e.g., Petria 1987). One of the main problems with global supervenience is that if two worlds differ at all (even if only in the slightest respect) then their supervening properties could be entirely different according to supervenience, which is rather counterintuitive. On the other hand, a similar more "local" problem might arise for the two other versions of supervenience as well. Assume that mental properties of humans supervene on physical properties of brains, e.g., that pain supervenes of C-fiber stimulation. Suppose D-fibers are very similar to C-fibers, in fact, they are functionally equivalent at a neural level, but do not have

---

[1] There are various versions of these three notions depending on how the modalities are defined. Furthermore, there are other kinds of supervenience, which involve relationships between patterns, or mathematical structures, e.g., one mathematical structure modeled in another.

[2] Note that the difference between strong and weak supervenience lies in the added necessity operator in the last conjunct.

certain low-level physical properties regarding their cell structure, etc. Then D-fibers could have entirely different supervening properties and this would be consistent with weak as well as strong supervenience. From a functionalist point of view, however, the two kinds of fibers should have the same supervening properties, if they have the same functionality (at a relevant level of description). This goes to show that supervenience is not quite sufficient for the functionalist to explain how the mental is determined by the physical if it is possible that the same functional role could lead to different mental properties. Consequently, none of the three versions of supervenience seem to be sufficient to capture the functionalist view on how the mental relates to the physical, i.e., how the mental is realized or implemented in the physical.

7.2.2 Relation Supervenience

There are other problems with a notion of supervenience that only considers the relation between sets of properties. It has been argued, for example, that relational mental properties do not supervene on properties of brains or organisms alone, but on environmental properties in addition (e.g., see Papineau 1995, or Heil 1995). While it seems that this deficiency could be remedied by adding environmental properties, it still leaves a strange aftertaste that something important about the relation of individual and environment is lost in such a move. But one does not even have to include the environment to see that reducing relational properties (if possible at all) will have unwanted side effects. One most obvious implication of such a restriction is that the language used to talk about mental architectures will be unnecessarily complicated if it has to be phrased it in terms of properties. For example, to say that "short term memory is connected to long term memory via recurrent excitatory connections" would translate the ternary relation "___connected to ___ via ___" into something like the property of "having a short term memory and a long term memory such that these memories are connected by recurrent connections." Of course, in the property reading it is not possible to "access" the parts of the expression that makes up the property definition, such as "recurrent connection," "connected by," "short term memory," etc., (One would have to introduce another property for all these terms...), hence it is not possible to state explicitly that the property really expresses a relation. Not only is this very clumsy, but it is also not clear how many mental properties one would have to introduce to capture all the various theoretically interesting relationships among them.

Another complicating factor is that some of these properties might not supervene even weakly, as they might be entirely dependent on internal configurations of the mental architecture at a given time (e.g., the recurrent memory connection might be achieved by many different physical connections at different times, and the physical property that gave rise to the recurrent connection at some point, might give rise to something else at a different time).

It is not clear that all mental relations can be couched in terms of physical properties and their relation. Take, for example, Kim's suggestion of how to define the supervenience of a mental relation R on a set of physical properties B (Kim 1993, p. 161):

> For any n-tuples, $\langle x_1, \ldots, x_n \rangle$ and $\langle y_1, \ldots, y_n \rangle$, if they are indiscernible in set B, then $R(x_1, \ldots, x_n)$ if and only if $R(y_1, \ldots, y_n)$, where $\langle x_1, \ldots, x_n \rangle$ and $\langle y_1, \ldots, y_n \rangle$ are in indiscernible in B just in case x_i is indiscernible from y_i in respect of B-properties.

As Kim notes, while this might be sufficient for relations like "taller than," which hold of tuples because of their "intrinsic" properties alone, it will not suffice in general; just consider causal relations such as "earlier than" or "east of." In the latter case, the notion of indiscernability will have to be defined in a different way. Kim's suggestion for such an account of indiscernability, in case an n-ary relation R is present in the base set, is as follows:

> Two entities x and y are indiscernable with respect to R if and only if for all $x_1, \ldots, x_{n-1}$ and $y_1, \ldots, y_{n-1}$, $R(x, x_1, \ldots, x_{n-1})$ if and only if $R(y, y_1, \ldots, y_{n-1})$ and $R(x_1, x, x_2, \ldots, x_{n-1})$ if and only if $R(y_1, y, y_2, \ldots, y_{n-1})$ and ... and $R(x_1, \ldots, x_{n-1}, x)$ if and only if $R(y_1, \ldots, y_{n-1}, y)$.

There are, however, problems with this requirement, as Kim points out, for suppose "we want to discuss whether a certain property P of wholes supervenes on the properties and relations characterizing their parts. Let X and Y be two distinct wholes with no overlapping parts, and suppose X consists of parts $x_1, \ldots, x_n$ and Y consists of $y_1, \ldots, y_m$. We would expect some properties of X and Y to depend on the relationships characterizing their parts–how these parts are organized and structured–as well as the properties of the parts [...]. What should we say about the conditions under X and Y may be said to be "mereological indiscernible"–that is, alike in respect of the way they are made up of parts? In a situation of this kind it would be absurd to enforce [the above requirement for indiscernability]. For suppose that a dyadic relation, R, holds for two mereological parts of X, $\langle x_1, x_2 \rangle$; [the above requirement for indiscernability] would require mereological indiscernability of X and Y that some y_j be related to x_2! Obviously, what we want is that X and Y be characterized by *the same relational structure*. [...] If x_1 has R to x_2, the same corresponding element of, y_j, of Y must have R to an appropriate y_h, not to x_2." (Kim 1993, p. 164)

Kim then excludes isomorphism between X and Y as too strong a condition for mereological indiscernability, and suggests that "we may do well to work with similarity in the subvenient base set, rather than insist on indiscernability, when relations are present. In particular, the supervenience of the properties of wholes might be more appropriately explained in terms of their *mereological similarity* rather than their *mereological indiscernability*." (Kim 1993, p. 164). Unfortunately, Kim then draws the conclusion that "strict relational supervenience–that is, relational supervenience satisfying [the above requirement for indiscernability]–may not be such a useful concept after all," mainly because of the problems with the above requirement for indiscernability, it seems. This, however, can be remedied, as we will show later, by using an extension of the notion of "bisimilarity" which underwrites the notion of implementation.

7.3 Virtual Machines and Virtual Machine Functionalism

In a sense, mereological supervenience is a special case of the general case of the supervenience of one ontology on another, namely the case where "higher level entities" can be seen as being composed of "lower level entities." The stratified view of our physical world is an example: molecules are made out of atoms, cells consist of many molecules, (higher) organisms are assemblies of many cells, etc. But while this mereological relationship seems to be generally true of "physical levels of description" (unless one wants to count some quantum decomposition phenomena as contradicting this claim), it is not true of all levels of descriptions, in particular, not of so-called implementation levels, i.e., levels of abstraction at which *virtual machines* are described.

The notion of "virtual machine" is prominent in cognitive science and even more so in computer science. By "virtual machine" we mean a specification of both an architecture and the kinds of processes that this architecture supports. The architecture specification typically includes basic entities or "parts" and their functional properties as well as descriptions of the various ways how these parts can be connected. Furthermore, it includes a specification of input and output kinds of the architecture and of operations that can be performed on them. If the virtual machine is a computational virtual machine, then it is convenient to think of its processes as being specified by some program, but there are also other ways of specifying interactions of various parts.

Example 1 When we talk about SCHEME being a virtual machine, for example, we certainly do not mean that it is indeed some sort of "machine" (in a standard sense of "machine")–compare this to the term "search engine," etc. Rather, there seem to be two parts involved, (1) the *programming language* SCHEME, which consists of symbols and compositions of these symbols and (2) the SCHEME interpreter (i.e., the semantics of SCHEME), which evaluates SCHEME expressions. The later is what is considered the "virtual machine," as it does *perform* an operation, namely that of interpreting SCHEME expressions. To do this, it needs additional primitives and objects (such as environments, etc., most of which are accessible from within SCHEME) that allow it to store and retrieve, to evaluate and modify SCHEME expressions, and consequently (as implied by "store and retrieve," etc.) it needs some sort of scratch space ("working memory") and storage space ("long term memory") to operate with (tokens of) SCHEME expressions.

Virtual machine functionalism, then, is the view that all mental states (and this includes "phenomenal states") are states of processes of a virtual machine and mental concepts can be understood in terms of the concepts defining the respective virtual machine architecture. Virtual machine functionalism differs from other functionalist accounts in that it distinguishes between structural features of an architecture (such as its parts, the states that they can be in, and how they are related to other parts) and the processes supported by it (e.g., all possible processes of one subsystem and how they can influence the behavior of the processes of another subsystem). While parts of the virtual machine are generally viewed as serving particular functional roles in the overall architecture specification (analogous to functionalist accounts of functional states), states of the virtual machine, i.e., states of the processes "running on it," are taken to instantiate mental states.

If minds are then viewed as certain kinds of virtual machines, as we suggest, then it follows that supervenience, in order to be of any interest in addressing the mind–body problem (or the problem of how virtual machines relate to physical machines or other virtual machines), must be a relation between ontologies, i.e., it must include more than properties and relations. In particular, there will be a complex relation between the entities and concepts used in the specification of the supervenient virtual machine A and the entities and concepts used in the subvenient virtual machine or physical system B: while A entities will be related to B entities or conglomerations thereof, it might not be possible to relate A concepts to B concepts. Put differently, while A entities are *implemented* or *realized* in B entities, A concepts might not be reducible to B concepts. To couch this in terms of the mind–body problem: virtual machine functionalism suggests that mental entities are implemented in physical entities, while mental concepts are not reducible to physical concepts. This way the dependence and determination of the supervening virtual machine on the subvenient or implementing virtual machine is guaranteed without being reductionist about the supervenient ontology.

Even if one does not want to subscribe to the "virtual machine functionalist" stance on mind, the underlying notion of supervenience of an ontology is still more basic and general than any restricted notion of property or even relation supervenience. Any adherent of the latter notions will still have to explain why "being a belief," for example, is a *property* of a physical object. First of all, it seems that "physics" *prima facie* does not supply the right kinds of objects that can have beliefs. If it does, then this needs to be pointed out. Furthermore, if my beliefs are inconsistent, for example, there does not have to be anything in the physical world that has that property of inconsistency and also has some physical property on which being a set of inconsistent beliefs supervenes. The burden of proof as to why such mental entities and their properties have to be reducible to physical properties is on the side of the adherent of property supervenience. And, furthermore, even if this difficulty could be resolved, an even trickier issue remains: as Kim (1993) points out at numerous places, "mind–body supervenience [...] does not state a solution to the mind–body problem; rather it states the problem itself" (ibid., p. 168). Rather, he believes that in order to obtain a substantive mind–body theory the dependence underlying the mind–body property covariance needs to be explicated and that mereological supervenience is a promising candidate. We agree with Kim that supervenience claims alone are insufficient to *explain* the mind–body dependence as they fail to show *how* the mind depends on the body. Any plausible candidate mind–body theory needs to explicate and explain the relationship between two ontologies, the mental and the physical, and one of the most promising candidates showing how the mental can be related to the physical is the virtual machine functionalism together with an appropriate notion of implementation.

7.4 Towards a Formal Specification of VM Architectures

The notion of "virtual machine architecture" or, more generally, "functional architecture" is central to the fields of cognitive science and artificial intelligence. It is thought to capture the basic information processing capabilities of natural or artificial information processing systems by specifying how functional "primitives" (or units), which cannot be explained in terms of decomposing them into smaller functional units, are related to each other to form a network of functions, all of which together define the information processing (and possibly cognitive) system. One of the interesting features of functional architectures is that they offer an escape from Ryle's Regress (sometimes also called "the homunculus fallacy") by using smallest non-decomposable functional units (these units are themselves explained by appealing to the properties of the systems implementing them).

Functional architectures can be used to explain mental states in terms of their functional (or causal) role, which can be decomposed into simpler terms (or states) until the smallest functional units are reached. If functionalism is right that mental states are defined solely by their causal role, then a functional specification of a cognitive architecture (i.e., providing a "functional architecture") is sufficient to study and explain mental phenomena at a certain level of description.

Some people have even linked functional specifications to defining a programming language: "Specifying the functional architecture of a system is like providing a manual that defines some programming language. Indeed, defining a programming language is equivalent to specifying the functional architecture of a virtual machine" (Pylyshyn 1984, p. 92).

In this section, we will work towards the definition of the notion of "functional unit," which is intended to capture the intuition that functional architectures consist of many different, yet connected functional parts or "subarchitectures," each of which can be in many different internal states. The basic idea of a "functional unit" is that it consists of input, inner, and output states as well as a transition function relating input and inner to output and inner states in time. Functional units will be allowed multiple inputs and outputs along what will be called "input or output channels". The notion of "channel" is to be understood in the sense of Shannon (1975) from an implementation point of view. From a logical point of view, however, it rather corresponds to the notion of "information channel," for example, as in Barwise and Seligman (1998).

7.4.1 Functional Units

Before tackling the complex case with multiple inputs and output, however, we shall start with the simple case of a functional unit with only one input and only one output. In a sense, such a functional unit is nothing but a finite state automaton without start state and final states, where transitions are labelled with a *duration*, i.e., the time it takes to transit from one state to the other. We will assume a (not necessarily finite) set Time, which comprises possible durations (e.g., "1 msec," "2 msec," etc.) and is closed under addition (i.e., any two durations can be added and will yield another duration that is again in Time. Note that no assumption is made as to *how* durations are specified, measured, etc. They are simply assumed to be given in advance. Also, there are issues regarding the nature of these durations that will not be addressed here (e.g., whether they are average durations).

Definition 1 A simple functional unit SFU is a tuple $\langle\langle Input\rangle, \langle Output\rangle, \langle Inner\rangle,$ $trans\rangle$ where $Input$ is the set of input states, $Inner$ is the set of inner state, $Output$ is the set of output states, and $trans$ is the transition function $trans : Input \times Inner \longrightarrow$ Time $\times$ $Output \times Inner$.

A special case is a functional unit with *no inner states*, where input states are directly related to output states. Such functional units will be called "atomic" (or "primitive" in Cummins' terms) and will become important later as basic building blocks in functional architectures.[3]

Definition 2 A simple atomic functional unit $SAFU$ is a tuple $\langle\langle Input\rangle, \langle Output\rangle,$ $trans\rangle$ where $Input$ is the set of input states, $Output$ is the set of output states, and $trans : Input \longrightarrow$ Time$\times Output$ the function mapping inputs to durations and outputs.

Obviously, a simple atomic functional unit is a functional unit in the sense that it can be written as $\langle Input, Output, \oslash, trans'\rangle$, where $trans' = \{\langle\langle a\rangle, \langle t, b\rangle\rangle |$ $\langle a, \langle t, b\rangle\rangle \in trans\}$.

Everything said so far can be extended to functional units that have multiple input and multiple output channels. Instead of talking about "the input set" or "the output set," we simply consider finite sequences of such sets.

Definition 3 A functional unit FU is a tuple $\langle Input, Output, Inner, trans\rangle$ where $Input$ is a finite sequence (denoted as tuple) of sets of input states, $Inner$ a finite sequence of sets of inner states, $Output$ a finite sequence of sets of output states, and $trans$ is the transition function defined on sequences of sets as $trans : Input \times Inner \longrightarrow$ Time $\times$ $Output \times Inner$.

An atomic functional unit AFU is a tuple $\langle Input, Output, trans\rangle$ where $Input$ is a finite sequence of input states, $Output$ is a finite sequence of output states, and $trans : Input \longrightarrow$ Time$\times Output$ the function mapping input sequences to durations and output sequences.

Note that this definition is very similar to what Chalmers (1996) has called "combinatorial state automaton" (or CSA, for short). One minor difference is that CSAs can be infinite in that they can have an infinite sequence of inner states, which functional

---

[3] Ideally, a functional specification should be able to reduce the overall functional architecture to atomic functional units and their connections.

units cannot have. Another more important difference is that they incorporate the duration of their state transitions explicitly (and thus timing constraints implicitly).

Eventually, we want to start building functional units from atomic functional units, i.e., from units of which we only know the IO-function, but where no inner states or state descriptions are available.[4] In this case, the connecting states, i.e., the output states of the first and input states of the second atomic functional unit will become the "inner states" of the new functional unit obtained by composition.

7.4.2 Composition of Functional Units

The next step is to use functional units to build more complex functional units. This can be achieved by connecting output channels of one functional unit to input channels of another (or possibly the same) functional unit. This connection will be effected by a "transducer function," which maps all possible values of an output channel in a 1–1 correspondence onto all possible values of an input channel.

Definition 4 (*Composition of Functional Units*) Let FU_1 and FU_2 be two functional units (not necessarily distinct) and let $Output_{FU_1}(i)$ be the set of values of the i-th output channel of FU_1 and $Input_{FU_2}(j)$ the set of values of the j-th input channel of FU_2. Then a bijective function f from $Output_{FU_1}(i)$ to $Input_{FU_2}(j)$ is called an "i, j-transducer" from FU_1 to FU_2. We then say that two functional units FU_1 and FU_2 are *i, j-composable* if there is an i, j-transducer from FU_1 to FU_2. And two functional units are said to be *composable* if they are i, j-composable for some i and j.

Transducers play an important role in connecting functional units; yet, they are mathematical constructions, and in the implementation of functional units links between units might have to be affected by a physical transducer. Hence, there are at least three different readings of "exists a bijection f": the general mathematical sense, the physical sense, and the practical sense, corresponding to three notions of possibility, the logical, physical, and feasible. For now, we will focus solely on the mathematical reading.

Merely connecting some input and output channels of functional units is not sufficient if we want to look at the resultant unit as a functional unit itself (as we have not defined what the "inner states" of the newly obtained unit are supposed to be). The idea then is to use a "product" of the inner states of both functional units in the sense that the new functional unit can at most have $n \cdot m$ different inner states, if the the first FU has n inner states and the second m.

Definition 5 Let FU_1 and FU_2 be two i,j-composable functional units and let f be an i,j-transducer from FU_1 to FU_2. Then the *f-composition* of FU_1 and FU_2 is

---

[4] Note that this does not imply that the unit does not have internal states of any internal structure, only that we do not know what the details of its inner states or internal structure.

the tuple $\langle Input_{FU_1}, Output_{FU_2}, Inner_{FU_1} \times Inner_{FU_2}, trans \rangle$ where $trans$ is the relation defined as the set of all tuples $\langle \langle in_1, \langle s_1, s_2 \rangle \rangle, \langle t, out_2, \langle s_1', s_2' \rangle \rangle \rangle$ such that

1. $in_1 \in Input_{FU_1}$
2. $out_2 \in Output_{FU_2}$
3. $\langle \langle in_1, s_1 \rangle, \langle t_1, out_1, s_1' \rangle \rangle \in trans_{FU_1}$ for some t_1
4. $\langle \langle in_2, s_2 \rangle, \langle t_2, out_2, s_2' \rangle \rangle \in trans_{FU_2}$ for some t_2
5. $f(out(i)_1) = in(j)_2$
6. $t = t_1 + t_2$[5]

The way the above definition stands is not quite satisfactory. First of all, it is not clear what is going to happen with the outputs of the first functional units that are not connected to any inputs of the second, and vice versa with the inputs of the second unit that are not connected to any outputs of the first. Are they not used? One way to remedy this is to add the remaining input channels of the second to the input channels of the first, and to do the same for the output channels. However, this might lead to unwanted effects because of the time difference at which the input signals arrive at the second functional unit: the inputs that go through the first functional unit are by a factor of t_1 delayed as opposed to the ones that go directly to the second unit, and thus the overall output might be different from a scenario, in which the corresponding inputs are all applied at the same time to the second unit (e.g., by adding a delay units to all the input channels of the second functional unit that are not connected to outputs of the first functional unit, and to all output channels of the first that are not connected to input channels of the second, as the output channels of the second functional unit are delayed by a factor of t_2).

By adding all input lines and all output lines, we arrive at the definition of a complete f-composition:

Definition 6 (*Complete f-composition*) Let FU_1 and FU_2 be two i, j-composable functional units and let f be an i,j-transducer from FU_1 to FU_2. Furthermore, let $Input$ be the concatenation of the sequences $Input_{FU_1}$ and $Input_{FU_2}/j$ (which is the sequence reduced by the j-th component) and $Output$ be the concatenation of the sequences $Output_{FU_1}/i$ (the sequence reduced by the i-th component) and $Output_{FU_2}$.

Then the *complete f-composition* $FU_1 \Rightarrow_f FU_2$ of FU_1 and FU_2 is the tuple $\langle Input, Output, Inner_{FU_1} \times Inner_{FU_2}, trans \rangle$ where $trans$ is the relation defined as the set of all tuples $\langle \langle in_1, \langle s_1, s_2 \rangle \rangle, \langle t, out_2, \langle s_1', s_2' \rangle \rangle \rangle$ such that

1. $in_1 \in Input$
2. $out_2 \in Output$
3. $\langle \langle in_1, s_1 \rangle, \langle t_1, out_1, s_1' \rangle \rangle \in trans_{FU_1}$ for some t_1
4. $\langle \langle in_2, s_2 \rangle, \langle t_2, out_2, s_2' \rangle \rangle \in trans_{FU_2}$ for some t_2

---

[5] This is the overall duration between any input and the propagated effect through the one connected line to any output.

5. $f(out(i)_1) = in(j)_2$
6. $t = t_1$ for $in_1 \in Input_{FU_1}$
7. $t = t_2$ for $in_1 \in Input_{FU_2}/j$

As a corollary we get that both ways of combining functional units result in new functional units (the only difference being that some of the input and output channels are not available if the combination is not complete).

Corollary 1 *The (complete) f-composition of two f-composable functional units is a functional unit.*

Proof Let $FU_1 \Rightarrow_f FU_2$ be the (complete) f-composition of two i,j-composable functional units FU_1 and FU_2 and let f be an i-j-transducer from FU_1 to FU_2. Obviously, $Input_{FU_1 \Rightarrow_f FU_2}$ $Inner_{FU_1 \Rightarrow_f FU_2}$ and $Output_{FU_1 \Rightarrow_f FU_2}$ are sets as required; to see that $trans_{FU_1 \Rightarrow_f FU_2}$ is of the right kind, observe that each tuple is of the form $\langle\langle in, \langle s_1, s_2\rangle\rangle, \langle t, out, \langle s_1', s_2'\rangle\rangle\rangle$, where $\langle s_1, s_2\rangle$ and $\langle s_1', s_2'\rangle$ are in $Inner_{FU_1 \Rightarrow_f FU_2}$ (as $Inner_{FU_1 \Rightarrow_f FU_2}$ is defined on $Inner_1 \times Inner_2$) and $t \in$ Time given that Time is closed under addition.

So f-composition can be used to create more complex functional units from simpler ones. Note that timing (as mediated by "duration") plays a crucial role in constructing more complex functional units: without the explicit duration parameter it would not be possible to distinguish circuits that can be distinguished now (e.g., and XOR gate with a self-feedback loop).

Example 2 Suppose we are given the two functional units $FU_1 = \{\langle\{b\}\rangle, \langle\{c\}\rangle, \langle\{S\}\rangle, \{\langle\langle b, S\rangle, \langle 15, c, S\rangle\rangle\}\}$, and $FU_2 = \{\langle\{a\}\rangle, \langle\{e, o\}\rangle, \langle\{O, E\}\rangle, \{\langle\langle a, E\rangle, \langle 10, o, O\rangle\rangle, \langle\langle a, O\rangle, \langle 10, e, E\rangle\rangle\}\}$ and the function $f = \{\langle c, a\rangle\}$ (Time is the set of positive integers). Observe that both units are 1,1-composable and that f is a 1,1-transducer from FU_1 to FU_2. Hence, the complete f-composition is $FU_1 \Rightarrow FU_2 = \{\{b\}, \{e, o\}, \{\langle S, E\rangle, \langle S, O\rangle\}, \{\langle\langle b, \langle S, E\rangle\rangle, \langle 25, o, \langle S, O\rangle\rangle\rangle, \langle\langle b, \langle S, O\rangle\rangle, \langle 25, e, \langle S, E\rangle\rangle\rangle\}\}$.

We also introduce another way of combining functional units which, different from the above combination, explicitly introduces a *new internal state*, namely the state of connection between the functional units as affected by the transducer. We will call "(complete) f-extension" to indicate a new internal state was that added to the functional unit.[6]

Definition 7 *(Complete f-extension)* Let FU_1 and FU_2 be two i,j-composable functional units and let f be an i, j-transducer from FU_1 to FU_2. Furthermore, let *Input* be the concatenation of the sequences $Input_{FU_1}$ and $Input_{FU_2}/j$ (which is the sequence reduced by the j-th component) and *Output* be the concatenation of the sequences $Output_{FU_1}/i$ (the sequence reduced by the i-th component) and $Output_{FU_2}$.

---

[6] Note that we can get an "incomplete" extension the same we got an incomplete composition by ignoring the input states of the second unit and the output states of the first that are not connected.

Then the *complete f-extension* $FU_1 \Rightarrow_f^+ FU_2$ of FU_1 and FU_2 is the tuple $\langle Input, Output, Inner_{FU_1} \times f \times Inner_{FU_2}, trans \rangle$ where *trans* is the relation defined as the set of all tuples $\langle \langle in_1, \langle s_1, n, s_2 \rangle \rangle, \langle t, out_2, \langle s_1', n', s_2' \rangle \rangle \rangle$ such that

1. $in_1 \in Input$
2. $out_2 \in Output$
3. $\langle \langle in_1, s_1 \rangle, \langle t_1, out_1, s_1' \rangle \rangle \in trans_{FU_1}$ for some t_1
4. $\langle \langle in_2, s_2 \rangle, \langle t_2, out_2, s_2' \rangle \rangle \in trans_{FU_2}$ for some t_2
5. $n, n' \in f(out_1(i)) = in_2(j)$
6. $t = t_1$ for $in_1 \in Input_{FU_1}$
7. $t = t_2$ for $in_1 \in Input_{FU_2}/j$

7.4.3 Functional Architectures and Their Realization

Now that we have a way for functional units to be composed and extended, we also need to define a criterion for when we consider them "functionally equivalent," i.e., when they produce the same input–output mappings for all input patterns and times. Moreover, since we are interested in internal states as well, we want to have a way of distinguishing functionally equivalent units that also have, in some sense, the same internal states from those that might have a different internal state structure. This is, for example, important for the kinds of higher level internal states that such a functional unit might realize (e.g., which mental states a particular functional architecture can instantiate). The idea here is based on Scheutz (2001) where we consider to computational systems the *same* if the internal states can be related through "bisimulation." Here we adapt the notion of bisimulation to functional units:

Definition 8 (*Bisimulation between Functional Units*) Let $FU_1 = \langle Input, Output, Inner_1, trans_1 \rangle$ and $FU_2 = \langle Input, Output, Inner_2, trans_2 \rangle$ be two functional units where *Input* and *Output* are two finite sequences of sets of input and output states, respectively, $Inner_1$ and $Inner_2$ are two sequences of sets of inner states, and $trans_1$ and $trans_2$ are the transition functions defined on sequences of sets as $trans_1 : Input \times Inner_1 \longrightarrow \mathsf{Time} \times Output \times Inner_1$ and $trans_2 : Input \times Inner_2 \longrightarrow \mathsf{Time} \times Output \times Inner_2$. The two functional units are said to be *bisimilar* if there exists a nonempty relation R (called "bisimulation") defined on $Inner_1 \times Inner_2$ such that the following four conditions hold:

1. If $\langle s_1, s_2 \rangle \in R$ and $\langle \langle i, s_1 \rangle, \langle t, o, s_1' \rangle \rangle \in trans_1)$, then there exists $s_2' \in Inner_2$ such that $\langle s_1', s_2' \rangle \in R$ and $\langle \langle i, s_2 \rangle, \langle t, o, s_2' \rangle \rangle \in trans_2)$
2. If $\langle s_1, s_2 \rangle \in R$ and $\langle \langle i, s_2 \rangle, \langle t, o, s_2' \rangle \rangle \in trans_2)$, then there exists $s_1' \in Inner_1$ such that $\langle s_1', s_2' \rangle \in R$ and $\langle \langle i, s_1 \rangle, \langle t, o, s_1' \rangle \rangle \in trans_1)$
3. For every $s_1 \in Inner_1$ there exists a $s_2 \in Inner_2$ such that $\langle s_1, s_2 \rangle \in R$
4. For every $s_2 \in Inner_2$ there exists a $s_1 \in Inner_1$ such that $\langle s_1, s_2 \rangle \in R$

The bisimulation relation bins internal states in each functional unit and relates them such that "redundant" but different internal states are grouped together (into equivalence classes where each state in a class is, from the perspective of the bisimulation relation, indistinguishable from the other members). As a corollary we get that there is a smallest bisimilar functional unit which has the minimal number of internal states necessary to affect the requisite input–output mapping given the internal state structure (compare this to the notion of characteristic automaton in Scheutz 2001).

As mentioned in the beginning, functional architectures are made up of functional units (simple or complex), each of which serves a (causal) role in the overall architecture:

Definition 9 (*Functional Architecture*) A *functional architecture* is a tuple $\langle Arch,$ $Parts, Labels \rangle$ where $Arch$ is a (usually composite) functional unit, $Parts$ is a *finite* set of functional units which are *components* of $Arch$ (in that they, if composed appropriately, will become *bisimilar* to $Arch$), and $Label$ is a function assigning each functional unit in $Parts$ a unique label (which can be used, for example, to describe the causal role of the labeled part; or it could be labeled with a predicate expressing the mental property realized by the part, etc.) [7]

Note that $Parts$ can be empty (in case $Arch$ is not made up of any components, i.e., if it is an atomic functional unit), or it may contain different kinds of functional units. If it contains only simple atomic functional units, then the advantage is that all "inner states" in $Arch$ can be explained as and arise from connections between units in $Parts$. Even if a functional architecture is not specified by virtue of only simple functional units (without inner states), it is always possible to decompose any functional architecture into atomic functional units, i.e., the given architecture can be specified by an *equivalent architecture* such that the set $Parts$ consists solely of atomic functional units (this requires a straightforward extension of the notion of bisimulation between two functional units to functional architectures such that the two $Arch$ functional units are to be bisimilar, but not necessarily the functional units in $Parts$). This is argument is supported by the following decomposition theorem:

Theorem 1 (Decomposition) *Every functional unit with inner states can be decomposed into atomic functional units without inner states.*

Proof (Sketch) Let FU be the functional unit given by $\langle Input, Output, Inner,$ $trans \rangle$ where $Input$ is a finite sequence of sets of input states, $Inner$ is a finite sequence of sets of inner states, $Output$ is a finite sequence of sets of output states, and $trans$ is the transition function defined on sequences of sets as $trans : Input \times$ $Inner \longrightarrow$ Time $\times Output \times Inner$. Then define two i,j-composable atomic functional units $AFU_1 = \langle Input \cup Inner, Inner \cup Inner, trans_1 \rangle$ – the "state updater" – and $AFU_2 = \langle Input \cup Inner, Output, trans_2 \rangle$ – the "output producer" – where $trans_1 : Input \times Inner \longrightarrow$ Time $\times Inner \times Inner$ and $trans_2 : Input \times$

---

[7] Cp. this with Copeland (1996) notion of architecture, which is based on the idea that one can label parts of a physical system, and with Gandy 1980, who uses hereditarily finite sets to form such hierarchies of parts of systems.

$Inner \longrightarrow \text{Time} \times Output$ such that $\langle\langle in, inner_m\rangle, \langle t, inner_n, inner_n\rangle\rangle \in trans_1$ and $\langle\langle in, inner_m\rangle, \langle t, out\rangle\rangle \in trans_2$ iff $\langle\langle in, inner_m\rangle, inner_n, out\rangle \in trans$. First, we recursively produce a new atomic functional unit $FU'_1 = \langle Input, Output, trans\rangle$ by recursively connecting all output lines for one set of inner states to the input lines for inner states using the identity function as a transducer (this functional unit now maps input states onto inner states of FU). Then recursive i,i-composition of the i-th output line from FU'_1 with the corresponding i-th input line for inner states for FU_2 for all i output lines of FU_1 (leaving only input lines for $Input$ states in FU_2) eventually leads to a functional unit that is bisimilar to FU. Finally, to use a new functional "split" unit which performs the function $split(X) = \langle X, X\rangle$ for all inputs sequences X, to duplicate the set of input channels in two and connect one set of output channels to FU'_1, and the other to FU_2.

It then follows that every functional architecture has an equivalent functional architecture consisting of only atomic functional units as parts and all internal states are the states on the connection lines. This is important because we already have a formal criterion for what it means to realize atomic functional units in something physical, say, based on the notion of "realization of a function" developed in Scheutz (1999).

Definition 10 (*Realization of Atomic Functional Unit*) An atomic functional unit $AFU = \langle Input, Output, trans\rangle$ with a finite sequence of input states $Input$, a finite sequence of output states $Output$, and a state transition function $trans$: $Input \longrightarrow \text{Time} \times Output$ mapping input sequences to durations and output sequences is realized by a physical system S (describable in a theory P) if and only if the following conditions hold:

1. There exists a (syntactic) isomorphic mapping I from the "input domain" of S to $Input$[8]
2. There exists a (syntactic) isomorphic mapping O from the "output domain" of S to $Output$
3. There exists a function F that describes the physical property (=behavior) of S for the given input-output properties (i.e., F is a mapping from the "input domain" of S to its "output domain" described in the language and by the laws of P) such that for all $x \in Input$, $O(F(I^{-1}(x))) = trans(x) \in Output$.

Given that we know what it means to realize the atomic functional units, we can provide at least one way to realize any functional unit by simply using the same construction as in the *Decomposition Theorem* and applying the above definition to the two decomposed units. And, as a result, we also can postulate at least one way of realizing a whole functional architecture, by realizing its *Arch* functional unit (according to the above definition), and by ensuring that the same physical system also realizes all atomic functional units. Of course, the "better way" of realizing a functional architecture in a physical system (or another functional architecture)

---

[8] See Scheutz (1999) about the qualifier "syntactic isomorphism" (for all practical purposes we can simply consider it here an isomorphic mapping).

is to ensure that all atomic functional units are realized in (distinct) parts of the physical system and that those physical parts are connected in the same way (i.e., the connection graphs among realized units in the two architectures are isomorphic) that the atomic functional units are connected to give rise to the whole functional architecture. This requires us also to check that the inputs and outputs of all connected physical parts match up (otherwise physical transducers will have to be introduced to ensure that parts can be connected properly). Moreover, we also have to pay attention to meet the timing constraints imposed by the parts of the functional architecture in the physical system (as different transition times might give rise to different functional systems). Note that the above construction not only works for physical systems, but also for other virtual machine architectures, thus allowing us to define a very general notion of one virtual machine being implemented in another virtual machine.

7.5 Discussion

The above sketch of how one could define a general notion of implementing one virtual machine in another by way of showing how their functional architectures can be related is, of course, only a start. For one, the notion of virtual machine used in the above discussion was very informal, and we attempted to make it more formal by introducing the notion of a functional unit as constituent part of a virtual machine. However, in this discussion, we glossed over the ontological status of input and output states, i.e., what kinds of entities those ranged over. For example, the inputs to one functional unit might be chess pieces, while the inputs and outputs to another might be polygons or words. As a result, there is an important open problem left to tackle, namely how to map entities onto each other when they do not belong to the same domain (e.g., real numbers). What is needed, effectively, is a theory of transduction and encoding that shows how entities of one kind can be obtained by putting together structures of entities of another kind (e.g., see Pollock (2008) for a promising approach). For example, in word processors implemented in von Neumann machines, words are made out of sequences of letters encoded in 7-bit binary ASCII binary codes. If one input line in a component of the word processing VM can hold any word (of at most 15 letters, say), this line would have to be mapped on 15*7 binary input lines in the implementing VM. Similarly, we might be able to encode some entities not only in terms of spatially separate codes, but temporally (e.g., in the sequential way the Morse code is realized). All of this will require more fleshing out of details in the above definitions to allow for more general mappings between different ontological entities. However, the overall structure of the relationship between components of an architecture, their connections, and the implementing structures will overall remain the same, i.e., the higher level VM that is to be implemented in a lower level VM will be defined by its architecture that either can or cannot be realized, in the above sense, in the architecture of the lower level VM. As a result, the notion of implementation of virtual machines in other virtual or physical machines show one way in which whole ontologies could be related, assuming that the problem of relating and encoding higher level into lower level entities can be defined in general enough terms (the details of this definition have to be left for another occasion). This also entails that *any relation* defined in terms of the states of connections among atomic functional units realized in one virtual machine are also realized in the implementing virtual machine. Moreover, the mapping of VM parts of the higher level VM onto VM parts in the lower level VM shows *how* the relation is realized. But note that the mapping does not mean that the *concepts* the intensions associated with those entities and relations *can be reduced*. In other words, VM implementation provides a way of showing how higher level entities can be encoded in, made out of, or related to possibly complex structures of low-level entities *without* forcing one to *equate* or *identify* higher level entities with those realizing structures, thus providing nonreductive bridges that preserve the conceptual autonomy of higher level concepts in the ontology without having to subscribe to spooky metaphysical theories (such as various forms of dualism) in order to explain the dependence of higher level on lower level virtual machines.

Acknowledgments This paper would not have been possible without the many discussions with Aaron Sloman the author was fortunate to have over the years, although Aaron is by no means to blame for any errors or potential problems with the specific content.

References

Barwise J, Seligman J (1998) Information flow. The logic of distributed systems. Cambridge Tracts in Theoretical Computer Science 44. Cambridge University Press, Cambridge

Chalmers D (1996) Does a rock implement every finite-state automaton? Synthese 108:309–333

Copeland J (1996) What is computation? Synthese 108:335–359

Gandy R (1980) Church's thesis and principles for mechanisms. Stud Log Found Math 101:123–148

Heil J (1995) Supervenience redux. In: Elias E. Savellos, Yalcin U (eds) Supervenience: new essays. Cambridge University Press

Kim J (1984) Concepts of supervenience. Philos Phenomel Res 14:153–176

Kim J (1993) Supervenience and mind: selected philosophical essays. Cambridge University Press, Cambridge

Kim J (1998) Mind in a physical world. MIT Press, Cambridge, Mass

Papineau D (1995) Arguments for supervenience and physical realization. In: Savellos E, Yalcin U (eds) Supervenience. Cambridge University Press, Cambridge

Petria B (1987) Global supervenience and reduction. Philos Phenomel Res 48:119–130

Pollock J (2008) What am i? virtual machines and the mind/body problem. Philos Phenomel Res 76:237–309

Pylyshyn Z (1984) Computation and cognition. MIT Press, Cambridge

Scheutz M (1999) When physical systems realize functions. Mind Mach 9:161–196

Scheutz M (2001) Causal versus computational complexity? Mind Mach 11:534–566

Shannon RE (1975) Systems simulation: the art and science. Prentice-Hall

Sloman A (1998) Supervenience and implementation. Technical Report, School of Computer Science, University of Birmingham

Chapter 8
Building for the Future: Architectures for the Next Generation of Intelligent Robots

Nick Hawes

8.1 Introduction

I first got interested in the study of architectures for intelligent systems as a naïve 20 year old.[1] When applying for a PhD position under Aaron's supervision I read a number of his papers related to the subject (probably including (Sloman 1994, 1998, 1999a, b; Wright et al. 1996)), and was almost instantly hooked on the idea that the behaviour of autonomous agents, from goldfish to humans, simulated sheepdogs to robots, could be codified and constructed using box diagrams (some of audacious complexity (Sloman 2003)). In the intervening years I have learnt that the route from boxes and lines to bits and limbs is not as simple or direct as I once imagined. However I am still convinced, perhaps more strongly than ever, of the power (descriptive, prescriptive, communicative) of the architectural approach to both understanding and creating intelligent systems.

Let us consider the challenge of creating an intelligent robot that could operate in a home or place or work, interact with the humans present and help them in their lives. Today this long-time dream of AI feels both closer and further away: closer because we can now witness impressive demonstrations of robots performing intelligent behaviour in home and office environments,[2] further away because large amount of task-, environment- and system-specific engineering, tweaking, parameterising and plain old last minute hacking that is required to make even the simplest

N. Hawes (✉)
School of Computer Science, University of Birmingham, Birmingham, UK
e-mail: n.a.hawes@cs.bham.ac.uk

[1] I will use the first person in this article where it seems appropriate given the personal nature of these reflections and their context in the symposium.

[2] See, for example, the amazing impact Willow Garage's PR2 is having across research groups in the US, or the impressive entrants in this year's RoboCup@Home competition.

J. L. Wyatt et al. (eds.), *From Animals to Robots and Back: Reflections on Hard Problems in the Study of Cognition*, Cognitive Systems Monographs 22, DOI: 10.1007/978-3-319-06614-1_8, © Springer International Publishing Switzerland 2014

demonstrations possible.[3] To put this in another way, we look closer to this long-term goal because of great progress made in the component technologies required to create such robot (mapping and localisation, vision, language processing, planning, etc.). We look further away because we lack the tools (in a broad sense) to integrate these components into complete systems that are able to operate beyond the confines of rigidly constrained environments and tasks.

That integration is hard is well understood in the robotics community. It was explicitly mentioned in the recent Strategic Research Agenda For Robotics In Europe as one of the major challenges facing the discipline (EUROP 2009). The problems of integration are also slowly being acknowledged by the AI community, although the fragmentation of AI into a large number of specialised subfields means that it is hard to get work where integration is the main contribution accepted to mainstream venues (although see the AAAI special tracks for example).

To address this issue of integration, we must start taking the systematic study of architecture seriously, as by their nature architectures support integration. We must study both the engineering constraints of the components that need to be integrated and the requirements of the problems we wish our systems to solve. We must make the goals and assumptions of our research explicit, and open our systems and results to the verification of the community. And we must, above all, stop reinventing the architectural ideas that already exist, although this can only happen once we have some agreement about what these ideas are and how they can be applied in practice. In this article, I support this point of view by discussing the current state of the art in architectures, and discussing how they relate to the practice of building systems. I then propose two approaches that can allow architectural ideas to have a meaningful impact on the community of researchers interested in constructing complete intelligent artefacts. Finally, I identify some *architectural design patterns* that tend to reoccur in intelligent systems, and discuss how the use of such patterns (rather than whole architectures) is one possible route towards the reuse of architectural ideas.

8.2 What Is an Architecture

Langley et al. (2008) define an architecture as "the underlying infrastructure for an intelligent system" which "includes those aspects of a cognitive agent that are constant over time and across different application domains." They also draw a direct analogy between the architecture of an intelligent system and the architecture of a building, where this is the permanent features ("foundation, roof, and rooms") as opposed to the parts that can be moved or replaced (such as "furniture and appliances"). Although viewing an architecture as something fixed provides a useful intuition, it is also potentially too restrictive: a system's architecture may change as the

---

[3] Just show up suitably early for any robotics demo, competition or project review to see what I mean. And I should add that I have spent many long days and nights in feverish demo prep.

system develops over time.[4] Aaron has argued that architectural change is necessary to account for the changes in competences of a developing human (Chappell and Sloman 2007), although it is arguable that artificial architectures have been free to ignore such change as they are a long way from capturing these levels of complexity and intelligence.

To tailor this definition more closely to the purpose of this article, it is worth exploring it further, in particular the notion of elements that are invariant across application domains. Our ultimate aim is to create a piece of (intelligent) software running on some suitable hardware. This software will be nothing more than a collection of implemented algorithms plus the information they pass between themselves (some of which will come from sensors and drive actuators). Following our previous definition, the architecture of the system will define parts of both the algorithms and information, plus how these parts interact.

Now imagine that your task is to create a series of different intelligent systems, each of which must solve related problems in a different domain. As a good software engineer you don't want to write your software from scratch for each system; however, you cannot use a single implementation to create every system because the inter-domain and -problem variation is too great. The sensible solution to this problem is to create a hierarchy of systems. At the top of this hierarchy is a system that includes only algorithms and information that can be used in every system (i.e. in every domain for at least one problem). By creating systems that inherit from this base system and then from these derived systems (as you would in class inheritance in object-orientated programming) you can create systems that get increasingly more specialised for their target domain and problems. Each inheritance step could add additional information (e.g. knowledge about a particular domain) or algorithms (to solve problems only present in the targeted subset of domains). Finally, when you have created a system capable of performing as required in a single domain you can start it up. At this point the algorithms and information that were already present in the system are joined by additional information that the system generates by sensing, processing and acting in it's world.

There are three important steps in this example. The first step is the design and implementation of the most general system that is appropriate for the collection of domains and problems you are interested in. The second step is the process of building on this general system to make additional more specific systems which focus on a subset of the domains and problems (ending with a final system ready for a given set of domains and problems). During this process the developer is gradually focusing on a more and specific region of *design space* (Sloman 1998). The third and final step is the process of creating a running version of this system.

The process of creating an architecture in this way is analogous to the process of creating a planner to solve a problem. In this process you start with a planner which you have designed to provide some general capabilities (e.g. support for some subset or superset of a version of PDDL (Edelkamp and Kissmann 2004)). Next you write

---

[4] It may be the case that such change is not noticeable over short periods, particularly when compared with the changes apparent to an architecture's contents.

a planning domain which allows the planner to solve some specific subset of all the possible problems it is capable of solving (analogous to adding domain knowledge and algorithms to an architecture). Finally, you must create a planning problem by instantiating the system with an initial state and goal defined in the language of the planning domain (analogous to actually starting the system based on the architecture). It is important to make these steps explicit because there are significant challenges at each step, and future architecture research should be explicit about which step it is addressing (particularly the relationship between the first two steps).

8.3 Architectures in Practice

When developing an intelligent system there are two different ways that an architecture design can be used: as a software template, or as a set of design rules. When using an architecture as a software template, developers are generally provided with a template system based on the architecture, and then they must fill in the gaps to produce a working system for their problem. These steps are directly analogous to the first two system design steps described above: the template provides the general system, then the developers add the domain-specific content. The practical advantage of using an architecture in this way is that some of the implementation work has already been done for you. The more theoretical advantage is that all systems created from the same toolkit are all direct realisations of the architectural design, allowing for some degree of cross-domain evaluation for the architecture (see (Hawes et al. 2007a) for more information on the problem of evaluating architectures). The disadvantage of using an architecture in this way is that you are constrained in what you can do, both by the software (e.g. what languages and functionalities it supports) and by the design (i.e. in the behaviours that are natural to implement in the architecture, cf. *structural bias* in (Dittes and Goerick 2011)).

When using an architecture as a set of design rules, developers generally take a written description of a design and implement it in the way they see fit. This may produce a system which has components that map directly to components in the architecture design, or a system that represents the architectural elements in more implicit ways (Hawes and Wyatt 2010). The architecture design becomes a useful vocabulary in the development process and generally constrains the overall design and implementation of the system, albeit in a less strict manner than when using a software template (Hawes et al. 2010). Accordingly, the advantages and disadvantages of using an architecture as a set of design rules are the converse of those of the software template: developers get increased flexibility in both design and implementation (as they can interpret the written description as liberally as they wish), at the expense of leveraging existing code and limiting the scope for cross-system evaluation.

Whilst the flexibility provided by the design rule approach to architectures is appealing (particularly when you don't yet know what kind of system you need to build), applying its practice leads to one of the major problems of cognitive systems work, both in theory and practice: the reinvention of ideas and technologies. This reinvention comes at a large cost in terms of both person months (as researchers hack away to integrate software components into complete system) and innovation (as researchers spend their time on ideas which have been previously explored by others). Such reinvention has stymied cognitive systems research, and will continue to do so until existing architectures (as both design and implementation) are routinely reused across our community. Therefore, the main message of this article is that researchers that concern themselves with architecture should strive to develop and disseminate their work in ways that encourage reuse. The following sections briefly outline two compatible ways in which this can be done, inspired by software engineering practices.

8.3.1 Software Toolkits

The first approach to fostering architectural reuse is to update the architecture template approach to match the software technologies available today. A large number of projects offer software frameworks for *component reuse* (e.g. (Fitzpatrick et al. 2008; Quigley et al. 2009)), but few offer complementary software (or even advice) for putting the component pieces together into an architecture (beyond component-to-component connectivity). Filling this gap, e.g. by producing a ROS stack[5] (Quigley et al. 2009) that provides all the necessary components and structure to quickly and easily implement a 3T-like architecture[6] (Bonasso et al. 1997) on a standard mobile platform, seems like a logical next step in the development of component-based frameworks. For such an approach to be successful, the software toolkit must also solve a collection of software-level problems that typically plague developers building complete systems from collections of components, in addition to providing high-level architectural structure. Such problems include dealing with concurrency, unpredictable delays in component execution and debugging complex heterogenous systems that interact with the real world. This is important because software frameworks will not be adopted if they make researchers' and developers' lives harder, and adoption is clearly coupled with reuse.[7]

8.3.2 Design Patterns

Although software toolkits provide a powerful, practical approach to architecture reuse, they also have the potential to be restrictive (in terms of where and how

---

[5] A collection of related software packages for the Robot Operating System.

[6] See Sect. 8.5.1.

[7] Related to this, it is worth noting that architecture toolkits seem to inspire a particularly strong strain of not-invented-here syndrome.

they can be used) and complex (as an architecture which supports a wide range of behaviours will have to have a large number of active elements). So rather than taking an all-or-nothing approach to architecture design reuse, we can follow the software engineering community and look for discrete design units that can be reused wherever appropriate in a complete system, i.e. *architectural design patterns*.[8] An example of such an architectural design pattern is the use of a planner coupled with a reactive system to execute its plans (see Sect. 8.5.1). Ideally, the use of any such pattern should be explicitly documented (in papers and source code), and be provided by explicit software support (much like the Model-View-Controller toolkits available for a large number of programming languages). The explicit nature of pattern use is important because it allows scientists and developers to study the pattern as an element on its own in many different systems, allowing them to build on, rather than reinvent, previous work. Some further suggestions for architectural design patterns are presented in Sect. 8.5. A closely related point was made previously by Bryson and Stein (2001). They promote the idea that new architecture features should be expressed in a commonly understood language such that the community can understand them with reference to existing features (also expressed in the same way), and thus compare them more easily and objectively. These *architecture idioms* are closely aligned with my design pattern suggestion.

8.4 A Potted Review

Although this article is not intended as a review of the field of architectures (many good reviews exist, including (Langley et al. 2008; Vernon et al. 2007)), it is instructive to sample what currently exists in light of the preceding sections (i.e. considering potential for (re)use creating intelligent robots).

Most research work that explicitly focusses on architectures for complete intelligent systems is performed with the aim of modelling, simulating or studying human cognition. This includes work on architectures such as ACT-R (Anderson et al. 2004), Clarion (Sun 2006), ICARUS (Langley and Choi 2006) and Soar (Laird et al. 1987), architectures which include a variety of learning and reasoning mechanisms coupled to predetermined logical representations. These architectures are particularly relevant to this article because they provide free software tools to allow researchers to develop systems using them. These software tools are (naturally) constrained to support only the architectures they are designed to implement, and thus have both the advantages (very good reuse and evaluation potential) and disadvantages (lack of flexibility) discussed previously. Domain- and task-specific content is typically added to these architectures by creating appropriate facts and rules in the supported production systems. This programming style is backed by a long tradition in AI, but

---

[8] Sometimes called *interaction patterns* in the *software* architecture literature, although these patterns focus typically on communication patterns between components, rather than larger functional units.

is not an ideal match for the distributed, concurrent, component-based style used to develop most intelligent robot systems. That said, these cognitive architectures are being increasingly applied to control robot systems and agents in virtual environments (e.g. (Choi et al. 2007; Benjamin et al. 2006)), and their software tools (if not their architecture designs) are consequently gaining associated features.

In terms of architectural reuse, these architectures do very well as their software toolkits generally provide a lot of preexisting structure and enough functionality to implement solutions to common problems. In the cases where toolkits are not available, then, as in all such cases, the reuse value is very small as developers often have to fill in much of the detail themselves.

One architecture where the absence of a software toolkit did not hinder reuse was the Subsumption architecture for reactive systems (Brooks 1986). Because its structure is simple and can be clearly explained in prose, it has been widely reused in both robots and other autonomous systems. However, the Subsumption architecture differs from the previously discussed architectures as it does not provide behavioural mechanisms, just constraints on component behaviours and interactions. An approach for designing architectures for reactive intelligence that does provide software tools and a development process is Bryson's Behaviour-Oriented Design (BOD) (Bryson 2003), which works with the POSH reactive planning system to create whole agents. For agents with purely reactive behaviour requirements, BOD and POSH provide an ideal solution for those interested in reuse and evaluation, as the software tools plus design methodology mean that a level of consistency can be achieved across different implementations.

Although BOD was originally designed for developing robots, it has seen most of its use for virtual characters. Developers of robot systems tend to prefer systems that fit the component-based manner in which they develop their systems (Fitzpatrick et al. 2008; Quigley et al. 2009). Architecture tools which fit this paradigm include PECAS (Hawes et al. 2009).[9] DIARC (Scheutz and Schermerhorn 2009), and Active Memory (Wrede 2008) Each of these provides a collection of software tools, rules for connecting components and functionality beyond component communication, and thus appear to have strong potential for both reuse and beneficial deployment on robots. However, to my knowledge none of these architectures are being used beyond the direct collaborators of the original developers (unlike some of the previously reviewed systems). One possible explanation for this is that all three systems use bespoke component systems which do not naturally interact with the component software frameworks used by most researchers, thus the many users of these extant frameworks would need to expend a lot of effort to reuse the architectural tools directly in their systems.

---

[9] This is the architecture I contributed to whilst building systems using CAST (Hawes and Wyatt 2010).

8.5 Design Patterns

Previously, I suggested that architectural design patterns are one approach that could be taken to promote the systematic and explicit reuse of architectural ideas in state-of-the-art robot systems. What follows is a list of patterns that seem necessary given systems currently being created. These are not derived from an analysis of the problems these systems are trying to solve (although they ultimately should be), but instead from seeing which problems are commonly solved by systems in the literature. As such they represent patterns which, if formalised further, could be put to immediate use by the community.

8.5.1 2-Layer Planning Architecture

Agents that plan their actions must have a system to take the resulting plans and execute them in the world. This is commonly done by using a reactive formalism (e.g. (Firby 1987; Nilsson 1994)) to actually implement each action from the planning domain. Examples of this can be found in a range of applications in the literature, e.g. (Bonasso et al. 1997; Hawes 2004; Malcolm 1997; Gat 1998; Talamadupula et al. 2010). There are many details of this integration of technologies that must be reinvented each time it is implemented (e.g. from the basic representations and communication patterns, to how to cope with action failure or interrupt), and each time this occurs an opportunity to build on previous experiences is lost. This integration of a planner and execution system has already almost made it to the level of a design pattern through its inclusion as two tiers (layers) in a three-tier planning architecture (usually the bottom two), e.g. 3T. See (Gat 1998) for a discussion of the three-tier architecture approach. As we are only interested here in the bottom two tiers (as they occur most commonly together in the literature), I refer to this proposed architectural design pattern as the *2T pattern*.

8.5.2 Anchoring

Agents that reason using abstractions over sensor data must have ways to transform incoming sensor data into these abstractions. This may involve creating new representations, updating and tracking them over time, and unifying different observations with a single existing representation as appropriate. This is usually referred to as *anchoring* (Coradeschi and Saffiotti 2003), and has been used to support temporal reasoning (Heintz et al. 2009), interaction (Fritsch et al. 2003) and planning (Hawes et al. 2007b). Whilst there are many different algorithms and approaches for the details of how the sensor data is treated (e.g. feature matching or bayesian approaches), it is arguable that the encompassing architecture could be invariant across these different approaches. This would again allow researchers to build on the experiences of others in designing and building such systems. I refer to this proposed architectural design pattern as the *anchoring pattern*.

8.5.3 Spatial Abstraction

Mobile agents that operate in large-scale space (i.e. space that extends beyond the range of their sensors (Kuipers 1988)) benefit from representing space at multiple levels of abstraction. Such abstractions may range from occupancy grids (Thrun et al. 2005) to metric SLAM maps (Folkesson et al. 2007) to (local) segmentations of space (e.g. navigation graphs (Sjöö et al. 2010), place representations (Pronobis et al. 2009)) to collections of these segmentations (area- and room-level topological maps (Hawes et al. 2011)). As with the anchoring pattern, the algorithms used to generate these representations from each other (and from sensor data) may vary wildly, but it is arguable that an architectural description of the process can be found that is invariant across different designs and implementations. This description may involve establishing dependencies between entities at different levels of abstraction, determining when to trigger updates to representations based on changes to others, and perhaps additions for propagating constraints from higher levels back down to lower levels. I refer to this proposed architectural design pattern as the *spatial abstraction pattern*. It is also possible that there are very similar patterns dealing with other abstraction hierarchies in other domains (e.g. the feature maps in visual attention (Itti and Koch 2000; Tsotsos et al. 1995)).

8.5.4 Putting Patterns to Work

The above pattens could all be implemented in software APIs that are representation and algorithm neutral (as far as possible). Researchers can then use these APIs to construct the architecture of their system (or parts of their system), before adding their own representations and algorithms into the architecture. This would be much like how component-based systems are used today (e.g. in ROS where a topic, e.g. laser_scan, can be published to by any component that can produce the correct message type), except that the system building would be occurring at a greater level of abstraction. In other words the researcher can assume that the architectural patterns determine how the components interact, and can concentrate on the component technologies (unless they are interested in the architecture itself).

It is the intention that the proposed architectural patterns can be combined into larger systems. For example entities from the spatial abstraction pattern could be anchored (via the anchoring pattern) to symbols which could be used for planning and execution in the 2T pattern. Such a combination would support the production of a basic deliberative architecture for a spatially embedded system (e.g. a mobile robot), by reusing standard architectural ideas as patterns.

8.6 Conclusion

In this article I have explored two ideas. The first is that the idea of architectures for intelligent systems is ripe for exploitation given the current state of component technologies and available software. The second idea is that in order to encourage progress in architecture research, we must concentrate on research methodologies that prevent us from continually reinventing and reimplementing existing work. The two ideas I propose for this are building software toolkits that provide useful architectures for the way researchers current develop systems, and focusing on architectural design patterns, rather than whole architectures. My ultimate aim in pushing this line of research is to be able to take ideas proposed and inspired by Aaron, those ideas that so enthused me a decade ago, and make them work for the researchers today.

References

Anderson JR, Bothell D, Byrne MD, Douglass S, Lebiere C, Qin Y (2004) An integrated theory of the mind. Psychol Rev 111(4):1036–1060

Benjamin DP, Lyons D, Lonsdale D (2006) Embodying a cognitive model in a mobile robot 6384(1):638,407. doi:10.1117/12.686163

Bonasso RP, Firby RJ, Gat E, Kortenkamp D, Miller DP, Slack MG (1997) Experiences with an architecture for intelligent, reactive agents. J Exp Theor Artif Intell 9(2–3):237–256

Brooks RA (1986) A robust layered control system for a mobile robot. IEEE J Robot Autom 2:14–23

Bryson JJ (2003) The behavior-oriented design of modular agent intelligence. In: Kowalszyk R, Müller JP, Tianfield H, Unland R (eds) Agent technologies, infrastructures, tools, and applications for e-services. Springer, Berlin, pp 61–76

Bryson JJ, Stein LA (2001) Architectures and idioms: making progress in agent design. In: Castelfranchi C, Lespérance Y (eds) The seventh international workshop on agent theories, architectures, and languages (ATAL2000), Springer, pp 73–88

Chappell J, Sloman A (2007) Natural and artificial meta-configured altricial information-processing systems. Int J Unconv Comput 3(3):211–239

Choi D, Könik T, Nejati N, Park C, Langley P (2007) A believable agent for first-person shooter games. In: Proceedings of the third artificial intelligence and interactive digital entertainment conference, pp 71–73

Coradeschi S, Saffiotti A (2003) An introduction to the anchoring problem. Robot Auton Syst 43(2–3):85–96. doi:10.1016/S0921-8890(03)00021-6,perceptualAnchoring: AnchoringSymbolstoSensorDatainSingleandMultipleRobotSystems

Dittes B, Goerick C (2011) A language for formal design of embedded intelligence research systems. Robot Auton Syst 59(3–4):181–193. doi:10.1016/j.robot.2011.01.001

Edelkamp S, Kissmann P (2004) PDDL 2.1: the language for the classical part of IPC-4. In: Proceedings of the international planning competition. international conference on automated planning and scheduling. Whistler, Canada

EUROP (2009) The strategic research agenda for robotics In Europe. http://www.robotics-platform.eu/sra

Firby RJ (1987) An investigation into reactive planning in complex domains. In: Proceedings of the Sixth national conference on artificial intelligence, pp 202–206

Fitzpatrick P, Metta G, Natale L (2008) Towards long-lived robot genes. Robot Auton Syst 56(1):29–45. http://dx.doi.org/10.1016/j.robot.2007.09.014

Folkesson J, Jensfelt P, Christensen HI (2007) The M-space feature representation for SLAM. IEEE Trans Robot 23(5):1024–1035

Fritsch J, Kleinehagenbrock M, Lang S, Plötz T, Fink GA, Sagerer G (2003) Multi-modal anchoring for human-robot interaction. Robot Auton Syst 43(2–3):133–147. doi:10.1016/S0921-8890(02)00355-X. In: Perceptual anchoring: anchoring symbols to sensor data in single and multiple robot systems

Gat E (1998) Three-layer architectures. Artificial intelligence and mobile robots: case studies of successful robot systems. MIT Press, Cambridge, pp 195–210

Hawes N (2004) Anytime deliberation for computer game agents. PhD thesis, School of Computer Science, University of Birmingham

Hawes N, Wyatt J (2010) Engineering intelligent information-processing systems with CAST. Adv Eng Inf 24(1):27–39. http://dx.doi.org/10.1016/j.aei.2009.08.010

Hawes N, Sloman A, Wyatt J (2007a) Towards an empirical exploration of design space. In: Kaminka GA, Burghart CR (eds) Evaluating architectures for intelligence: papers from the 2007 AAAI workshop. AAAI Press, Vancouver, pp 31–35

Hawes N, Sloman A, Wyatt J, Zillich M, Jacobsson H, Kruijff GJ, Brenner M, Berginc G, Skočaj D (2007b) Towards an integrated robot with multiple cognitive functions. In: Holte RC, Howe A (eds) Proceedings of the twenty-second AAAI conference on artificial intelligence (AAAI 2008). AAAI Press, Vancouver, Canada, pp 1548–1553

Hawes N, Brenner M, Sjöö K (2009) Planning as an architectural control mechanism. In: HRI '09: Proceedings of the 4th ACM/IEEE international conference on Human robot interaction, ACM, New York, NY, USA, pp 229–230, http://doi.acm.org/10.1145/1514095.1514150

Hawes N, Zillich M, Jensfelt P (2010) Lessons learnt from scenario-based integration. In: Christensen HI, Kruijff GJM, Wyatt JL (eds) Cognitive systems, cognitive systems monographs, vol 8. Springer, Berlin, pp 423–438

Hawes N, Hanheide M, Hargreaves J, Page B, Zender H, Jensfelt P (2011) Home alone: Autonomous extension and correction of spatial representations. In: Proceedings of the IEEE International Conference on Robotics and Automation (ICRA '11)

Heintz F, Kvarnström J, Doherty P (2009) A stream-based hierarchical anchoring framework. In: Proceedings of the 2009 IEEE/RSJ international conference on Intelligent robots and systems, IEEE Press, Piscataway, NJ, USA, IROS'09, pp 5254–5260

Itti L, Koch C (2000) A saliency-based search mechanism for overt and covert shifts of visual attention. Vis Res 40(10–12):1489–1506

Kuipers B (1988) Navigation and mapping in large-scale space. AI Mag 9:25–43

Laird JE, Newell A, Rosenbloom PS (1987) Soar: an architecture for general intelligence. Artif Intell 33(3):1–64

Langley P, Choi D (2006) A unified cognitive architecture for physical agents. In: Proceedings of the twenty-first national conference on artificial intelligence

Langley P, Laird JE, Rogers S (2008) Cognitive Architectures: Research Issues and Challenges. Cognit Syst Res 10(2):141–160. doi:10.1016/j.cogsys.2006.07.004

Malcolm C (1997) A hybrid behavioural/knowledge-based approach to robotic assembly. Evolutionary robotics: from intelligent robots to artificial life (ER'97). AAI Books, Tokyo, pp 221–256

Nilsson NJ (1994) Teleo-reactive programs for agent control. J Artif Intell Res 1:139–158

Pronobis A, Sjöö K, Aydemir A, Bishop AN, Jensfelt P (2009) A framework for robust cognitive spatial mapping. In: 10th international conference on advanced robotics (ICAR 2009)

Quigley M, Conley K, Gerkey B, Faust J, Foote T, Leibs J, Wheeler R, Ng AY (2009) ROS: an open-source robot operating system. In: ICRA workshop on open source software

Scheutz M, Schermerhorn P (2009) Affective goal and task selection for social robots. In: Casacuberta D, Vallverdú J (eds) The handbook of research on synthetic emotions and sociable robotics, Information Science Reference

Sjöö K, Zender H, Jensfelt P, Kruijff GJM, Pronobis A, Hawes N, Brenner M (2010) The explorer system. In: Christensen HI, Kruijff GJM, Wyatt JL (eds) Cognitive systems, cognitive systems monographs, vol 8. Springer, Berlin, pp 395–421

Sloman A (1994) Explorations in design space. In: Cohn A (ed) Proceedings 11th European conference on AI, Amsterdam, August 1994. Wiley, Chichester, pp 578–582

Sloman A (1998) The "semantics" of evolution: trajectories and trade-offs in design space and niche space. In: Coelho H (ed) Progress in artificial intelligence, 6th Iberoamerican conference on AI (IBERAMIA). Springer, Lecture Notes in artificial intelligence, Lisbon, pp 27–38

Sloman A (1999a) Beyond shallow models of emotion. In: Andre E (ed) Behaviour planning for life-like avatars, Sitges, Spain, pp 35–42, Proceedings I3 spring days workshop March 9th–10th 1999

Sloman A (1999b) What sort of architecture is required for a human-like agent? In: Wooldridge M, Rao A (eds) Foundations of rational agency. Kluwer Academic, Dordrecht, pp 35–52

Sloman A (2003) The cognition and affect project: architectures, architecture-schemas. School of Computer Science, University of Birmingham, And The New Science of Mind (Tech. rep)

Sun R (2006) The CLARION cognitive architecture: extending cognitive modeling to social simulation. In: Sun R (ed) Cognition and multi-agent interaction. Cambridge University Press, New York, pp 79–99

Talamadupula K, Benton J, Schermerhorn P, Kambhampati S, Scheutz M (2010) Integrating a closed world planner with an open world robot: a case study. In: AAAI conference on artificial intelligence

Thrun S, Burgard W, Fox D (2005) Probabilistic Robotics. MIT Press, Cambridge

Tsotsos JK, Culhane SM, Winky WYK, Lai Y, Davis N, Nuflo F (1995) Modeling visual attention via selective tuning. Artif Intell 78(1–2):507–545. doi:10.1016/0004-3702(95)00025-9

Vernon D, Metta G, Sandini G (2007) A survey of artificial cognitive systems: implications for the autonomous development of mental capabilities in computational agents. IEEE Trans Evol Comput 11(2):151–180

Wrede S (2008) An information-driven architecture for cognitive systems research. PhD thesis, Bielefeld University

Wright I, Sloman A, Beaudoin L (1996) Towards a design-based analysis of emotional episodes. Philos Psychiatry Psychol 3(2):101–126 (repr. in Chrisley RL (ed) (2000) Artificial intelligence: critical concepts in cognitive science, vol IV. Routledge, London)

Chapter 9
What Vision Can, Can't and Should Do

Michael Zillich

9.1 Introduction

Computer vision has made huge advances since its beginnings in the 1960s. After a slow start, plagued by limited computing power and sometimes overly optimistic predictions (such as implementing a generic vision system over a summer), recent years have seen increasing numbers of real world applications appearing on the market, from face tracking in consumer digital cameras, driving assistance systems in cars, autonomous vacuum cleaning robots to augmented reality applications or home entertainment. Of course industrial machine vision confined to the clearly structured environments of factory floors and assembly lines or medical imaging applications with a human in the loop have been on the market far longer. But within the scope of this article, we are interested in computer vision as it was seen by its early proponents as exemplified by Roberts (1965), Binford (1971), Clowes (1971), Huffman (1971), Waltz (1975), Nevatia and Binford (1977), Marr (1982), Biederman (1987): to understand the computational principles that allow human or animal vision to seemingly arrive at generic scene interpretations from images. Or put another way, vision that serves an agent to operate in and interact with the unconstrained real three-dimensional world.

This is of course a very broad definition and encompasses many different abilities related to locomotion, manipulation, learning, recognition, or social interactions, as has been emphasised by Sloman (1989). The many tasks that vision thus has to fulfil go beyond merely reconstructing and interpreting a three-dimensional scene. Many intermediate or very specific results of visual processing serve as input to, e.g. motor control or influence affective states. Part of the recent successes of vision, apart from increased computing power and the availability of new mathematical tools, is a high degree of specialisation of solutions in each of these areas.

M. Zillich (✉)
Vienna University of Technology, Vienna, Austria
e-mail: zillich@acin.tuwien.ac.at

J. L. Wyatt et al. (eds.), *From Animals to Robots and Back: Reflections on Hard Problems in the Study of Cognition*, Cognitive Systems Monographs 22, DOI: 10.1007/978-3-319-06614-1_9, © Springer International Publishing Switzerland 2014

For many of these specialised solutions impressive videos can be watched online and one is left wondering "Ok, done! So, where is the problem?". Yet performance of robots at competitions like the Semantic Robot Vision Challenge[1] or RoboCup@Home,[2] while clearly progressing from year to year, show that still these parts do not necessarily make a whole. There are of course many more problems to be solved in a complete robotic system besides vision, such as issues of power consumption and dexterity in manipulation, but limitations in perception and most notably vision typically do play a central role. So something must be still missing.

In the next section, we will review a selection of state-of-the-art solutions in some of the respective areas, showing which impressive things computer vision can in fact already do. Section 9.3 will then try to identify some fundamental problems in the current approach to computer vision, followed by suggestions on where future research could advance on a broader basis in Sect. 9.4.

9.2 What Vision Can Do

The following selection of work is not intended as a genuine review of work in different areas of computer vision, but rather to highlight some state-of-the-art solutions that taken together could seem to have solved computer vision for robotics. So, what can vision do for robotics?

9.2.1 Navigation, Localisation

Self localisation and mapping (SLAM) has been addressed by the robotics community early on, starting with ultrasonic and later laser range sensors where it can essentially be considered solved (Thrun et al. 2005). With increasing computational power and mathematical tools such as sparse bundle adjustment (Lourakis and Argyros 2009) vision based methods (Visual SLAM) began to replace laser based ones, e.g. (Nistér et al. 2006; Davison et al. 2007), which can now handle very large areas (Cummins and Newman 2010).

These methods rely on the robust extraction of uniquely identifiable image regions, for which a variety of image features have been proposed, such as MSER (Matas et al. 2002), SIFT (Lowe 2004), FAST (Rosten and Drummond 2006), SURF (Bay et al. 2008) or DAISY (Tola et al. 2008). These features, in general, play an essential role in many modern computer vision approaches, from SLAM and structure from motion to object recognition and tracking.

---

[1] http://www.semantic-robot-vision-challenge.org.

[2] http://www.ai.rug.nl/robocupathome.

9.2.2 3D Reconstruction

Using similar techniques, structure from motion (SfM) approaches put an emphasis of dense reconstruction of the scene rather than navigation based on sparse landmarks. Microsoft's Photo Tourism (Snavely et al. 2006) is quite well known. It can reconstruct in very high detail a building such as the cathedral of Notre Dame from a collection of thousands of photographs taken from the web. Going even further (Agarwal et al. 2009) scales the approach to entire cities, although that does take, as the title of their chapter suggests, a day of computation on a cluster of 500 computers.

Real-time solutions are available for smaller-scale scenes. The approach by Klein and Murray (2007) uses a parallel processing pipeline highly optimised to today's multi-core machines to build a semi-dense map of the environment based on tracking distinctive image features. Based on that the work by Newcombe and Davison (2010) fills in the details using GPU-based optical flow computation (Zach et al. 2007) to arrive at a dense 3D scene reconstruction with visually very pleasing results.

9.2.3 Scene Segmentation

The above approaches reconstruct the scene as a whole, essentially treating it as a single rigid and static object. Multibody structure from motion approaches, (Fitzgibbon and Zisserman 2000; Ozden et al. 2010) observe a dynamic scene and segment it into independently moving rigid objects.

Given only a static scene (Rusu et al. 2009) segments a 3D point cloud as provided by stereo or depth sensors into parametric object models such as planes, spheres, cylinders, and cones. Similarly Biegelbauer et al. (2010) fit superquadrics to point clouds to seamlessly cover a wider range of parametric shapes. Using a strong prior model of the 3D scene and again parametric object models Hager and Wegbreit (2011) is able to handle scenes exhibiting complex support and occlusion relations between objects, and also reasons explicitly about dynamic changes of the scene such as objects being moved, added or removed.

Taking a more active approach Björkman and Kragic (2010) combine wide angle and foveated stereo to segment 3D objects of arbitrary shape standing isolated on a supporting surface. Even more active, Fitzpatrick and Metta (2003) use a robot manipulator to poke parts of the scene in order to use the resulting motion in 2D image sequences to segment objects.

Recent advances in 3D sensing, most notably the Microsoft Kinect RBD-D sensor, brought a renewed interest in 3D methods. Having (close to) veridic depth perception simplifies the segmentation problem and allows segmentation of quite cluttered indoor scenes in real-time (Ückermann et al. 2012) into objects described in terms of parametric surface models (Richtsfeld et al. 2012).

9.2.4 Recognition

Object recognition is of course a central theme in computer vision especially in the context of robotics. Early attempts at generic recognition of 3D solids (Binford 1971; Waltz 1975; Nevatia and Binford 1977; Marr and Nishihara 1978; Brooks 1983; Biederman 1987; Lowe 1987; Dickinson et al. 1992), often based on edge features, tended to suffer from scene complexity and textured surfaces. With the advent of invariant interest point detectors (Mikolajczyk and Schmid 2004) and strongly distinctive point descriptors mentioned above (Matas et al. 2002; Lowe 2004; Rosten and Drummond 2006; Bay et al. 2008; Tola et al. 2008) appearance based recognition of arbitrarily shaped object instances in highly cluttered real world environments was essentially solved, e.g. (Lowe 1999; Gordon and Lowe 2006; Ferrari et al. 2006; Özuysal et al. 2007; Collet et al. 2009; Mörwald et al. 2010), even for non-rigid objects such as clothing (Pilet et al. 2007)—provided of course that the respective objects are textured. Making use a combination of colour image and dense depth map the fast template based approach by Hinterstoisser et al. (2011) also detects untextured objects in heavy clutter at close to frame rate.

The above appearance-based methods are intrinsically suited to detect individual object instances with specific surface markings. Going beyond single instances approaches (Fei-Fei et al. 2006; Leibe and Schiele 2003; Dalal and Triggs 2005) detect categories also of deformable objects such as cows or walking humans.

9.2.5 Online Learning

Acquiring models for the above recognition methods often involves hand-labelling of images or placing objects on turn tables as an offline learning step, which is clearly not desirable for an agent supposed to act autonomously in the world.

Various online learning methods have been proposed, such as Özuysal et al. (2006) which keeps "harvesting" additional features as it tracks the model acquired so far. The ProFORMA system (Pan et al. 2009) even reconstructs high quality dense triangle meshes while tracking a model and also suggesting new views to add.

Going further in the direction of a complete system Kraft et al. (2008) and Welke et al. (2010) let a robot pick up and rotate objects in its hand to actively cover all views of an object.

9.2.6 Tracking

Much as recognition, model-based 3D object tracking has been well covered in computer vision (Lepetit and Fua 2005). Especially with the availability of cheap and powerful graphics cards computationally heavy methods such as particle filtering (Klein and Murray 2006) have been rendered real time (Chestnutt et al. 2007; Murphy-Chutorian and Trivedi 2008; Sánchez et al. 2010; Choi and Christensen 2010; Mörwald et al. 2011) and allow tracking of complex 3D objects through heavy clutter.

So far for an (incomplete) overview of some of the success stories of computer vision in the realm of robotics. Next we will look at where we stand with this and why service robots are not yet scurrying around in our apartments.

9.3 What Vision Can't Do

What vision can't do is simply to allow a robot to operate in and interact with the unconstrained real three-dimensional world, as was our stated goal in the introduction.

9.3.1 Abstraction

One of the reasons why this is the case is explored in the very comprehensive review by Dickinson (2009). The author there sums up the evolution of object categorisation over the past four decades as different attempts to bridge the large representational gap between the raw input image at the lowest level of abstraction and 3D, viewpoint invariant, categorical shape models at the highest level of abstraction. In the 1970s, this gap was closed by using idealised images of textureless objects in controlled lighting to extract quite generic shape models. In the 1980s, the images could become more complex, by sacrificing model generality and searching for specific 3D shapes, thus effectively closing the gap at a lower level. Methods of the 1990's allowed recognition of complex textured objects in cluttered scenes, however objects were now essentially 2D appearance models of specific instances (and even views), thus closing the gap very low at the image level. The feature-based methods of the 2000s allowed recognition of arbitrary 3D object instances in very cluttered environments, while also slowly extending generality back up towards object categories.

So, much of the success of vision was bought by sacrificing generality and the ability of abstraction. This is less of a problem for navigation, where anything is an obstacle or a landmark, but more so for purposeful interaction with specific parts of the scene, viz, objects. Learning each object individually or perhaps narrow categories of objects is not feasible in the long run and does not provide the ability to make sense of a given scene even though a similar scene was never encountered. Humans, say an Inuit seeing tropical jungle for the first time, have no problem perceiving a completely unfamiliar scene in terms of complete 3D shapes plus their possibilities of interaction rather than an assortment of object and category labels. Otherwise the mentioned Inuit would have to appear essentially cortically blind, having no categories for all the different tropical trees and bushes.

9.3.2 Putting it Together

Another reason as to why the assorted successes of vision do not yet comprise a unified solution for robotic vision lies in the difficulties of merging these specialised solutions under one framework. One of the difficulties is robustness. Many methods rely on tuning of parameters or some hidden implicit assumptions. Operating these methods outside their safe zones can make them fail abruptly rather than degrade gracefully, leaving a system comprised of many such isolated solutions extremely brittle. A more severe problem actually lies in bridging the semantic gaps between different methods. What does it mean if an object recogniser reports a confidence of 0.4 of detecting an object right inside a wall while robot localisation reports an uncertainty of 40 cm? Approaches like Hoiem et al. (2006) have started exploring the interplay between, e.g. object recognition and estimation of coarse 3D scene geometry, and the work by Hager and Wegbreit (2011) mentioned above explicitly reasons about support and occlusion relations between objects. But a more generic solution of integrating the semantics (together with uncertainties) of individual processing results still seems far off.

9.3.3 Dealing with Failure

A third, somewhat related reason is that researchers in individual specialised sub-fields (quite naturally) strive for perfection, inching recognition rates on standardised benchmarks ever higher in increments of 0.5 %, while from a systems perspective it makes more sense to accept the inevitable uncertainties and failure modes and reason explicitly about them. This of course hinges on having a common framework as explained above, to meaningfully express these uncertainties. Even more importantly however it requires researchers to accept that perfection is futile.

9.4 What Vision Should Do

So what should be done to alleviate the above problems? There is of course no simple answer to this. But let us first look at some of the apparent solutions.

9.4.1 It Isn't 3D

With the availability of cheap and powerful 3D sensors such as complete stereo solutions by companies like Point Grey[3] or Videre Design[4] or depth sensors such as

---

[3] http://www.ptgrey.com/products/stereo.asp.

[4] http://www.videredesign.com.

the Mesa Imaging Swissranger[5] or Microsoft Kinect[6] one important part of vision seems to be have been solved, namely reconstructing a 3D scene. There is more to it though, as humans do not perceive a scene as a sort of flip-up cardboard diorama with missing object back sides. Reasoning about occluded parts of the scene as well as generic segmentation into individual objects remains to be solved.

More importantly however, a quick test on yourself by closing one eye will reveal that 3D sensing is not all that important for human vision. Picking up a small object or putting a key into a lock might require several attempts, so clearly direct perception of distance via stereo is an advantage for close-range manipulation, such as using tools or grasping branches when swinging from tree to tree. But the vivid impression of being situated in a 3D scene does not suffer significantly when being deprived of stereo vision. Also many grazing animals tend to have non-overlapping fields of view of the left and right eye, as large field of view (to notice approaching predators) is more important than accurate 3D perception.

For various reasons cups and cows are prominent example objects in computer vision. Advocates of 3D computer vision will point out that given a cup with a picture of a cow printed on it, a 2D recogniser would be likely to rather recognise a (nicely textured) cow than a (probably untextured) porcelain cup, whereas a 3D shape based recogniser would correctly identify the cup. However, given a 2D image of a cup with a cow on it humans have no problem recognising both, the cup and the fact that there is a picture of a cow printed on it.

We are not arguing that 3D sensing would not be a powerful cue, and in fact robotics is likely to benefit a lot from depth sensors in the near future. But 3D sensing does not seem to be essential for perceiving a 3D scene. Human vision has developed powerful computational mechanisms to infer a complete 3D scene from rather limited information. And these mechanisms are more important then a specific sensing modality.

9.4.2 It Isn't Resolution

In a similar vein image resolution does not seem to be critical. Certainly nature has evolved foveated vision for a good reason. The combination of attentional mechanisms based on low resolution cues with saccades to salient image regions to be processed at high resolution is an important mechanism to optimise visual processing and keep the amount of information tractable. Likewise any computer vision task benefits significantly from the object of interest being shown large and centred in the image, rather than occupying a small image region somewhere in the scene. However, humans looking at a low resolution, say 640×480, image of a scene typically have no problem interpreting it correctly (otherwise watching TV would be rather confusing).

---

[5] http://www.mesa-imaging.ch.

[6] http://www.xbox.com/en-US/kinect.

Moreover, experiments with rapid serial visual presentation (Thorpe and Imbert 1989) have shown that humans are remarkably good at identifying objects and scenes at presentation times well below 200 ms, which leaves no time to perform any saccades. For example in Intraub (1981) subjects were able to identify pictures of a category ('look for a butterfly'), superordinate category ('look for an animal') or negative category ('look for a picture that is *not* of house furnishings and decorations') presented for only 114 ms.

So the human visual system can perform (at least a significant deal of) its processing within an instant without requiring to scan the image with the high resolution fovea.

9.4.3 It Isn't (Just) Bayes

There is no doubt that much progress in vision is owed to the adoption of probabilistic frameworks over crisp symbolic methods, which are typically too brittle when confronted with the cluttered, uncertain, ambiguous real world. However sometimes the actual probabilities at the end of some lengthy mathematical argument are rather ad-hoc, say the number of matching edgels divided by the total number of edgels, or assumptions about uniform priors. The respective approaches still work fine, thanks to the extraordinary robustness of statistical methods. But the results from different processing modules, although supposedly derived within the same mathematical framework, become difficult to compare to each other within a common system. Just using probabilities is not enough. Care has to be taken, that they refer in the same way to the same underlying causes.

A different way to treat uncertainties, rather than aiming for precise 3D estimates plus a measure or remaining uncertainty, could be to not aim for exactness in the first place. Instead one could use more qualitative measures, such as surface A is behind surface B, which is an observation that can be established with high certainty over a wide range of actual distances. This sort of information might be sufficiently accurate for many types of actions, such as reaching for A. However, it is not clear whether the mathematics for this kind of reasoning over a complex 3D scene would actually turn out to be simpler than more traditional probability theory.

9.4.4 Back to the Roots?

Armed with the lessons learned along the way (and with considerably increased computing power), it might be worth reconsidering some of the early approaches to computer vision. These in general aimed at reconstructing a 3D scene from very impoverished visual information, such as edge images only. While simply taking a depth sensor would certainly provide a more direct route, attacking the harder problem, with all the modern mathematical machinery, is still worthwhile. Partly because advances there are more likely to shed light on the computational principles underlying human vision. But also because, as pointed out above, the problem of dealing with incomplete and ambiguous information persists, no matter how rich the underlying sensory information. This problem can be pushed back a little by more advanced sensors, but not avoided altogether.

9.4.5 A Conjecture: Vision as Prediction

In the following, we will put forward a conjecture of what might be one of the computational principles underlying human vision, based on an anecdotal example of severely impoverished visual information.

> I enter a room at night, put a glass of water on a table (Fig. 9.1a), walk back to the door to switch off the lights and the room becomes almost completely dark. I walk back to the table and can not actually see the glass or anything else on the table, or even the table itself. Scene reconstruction in this case is simply hopeless. Still I expect the glass at the same position where I left it and I can very roughly estimate that position by backtracking my steps. So I turn my head this way and that looking towards a window, which is slightly illuminated from outside, until I can see a glint typical of glass surfaces near the expected position (Fig. 9.1b). I reach out (carefully, as I might still collide with other unseen objects on the table) and successfully grab the glass (relying of course heavily on tactile feedback). By no means could I have reconstructed the glass with its 3D shape in that case, still I did 'see' it. Or rather I saw something I expected to see given that the glass were there.

Vision as a process of reconstructing the 3D scene is an ill-posed problem, yet humans seem to do it effortlessly. Still there are enough every-day cases where also for humans scene reconstruction becomes quite impossible (e.g. in very low light situations). Humans can however still employ vision successfully in such cases, because what if not vision eventually allowed the detection of the glass in the above-darkened room example. Only a little visual information was needed to confirm some hypothesis about the scene.

The point is that that while reconstruction is a notoriously difficult problem, the inverse problem—prediction—is often very simple. One avenue of progress might thus be to view vision (at least in part) as a prediction problem, based on strong priors. A general framework should be able to incorporate multiple cues (visual and possibly non-visual), where the appearance of each cue (such as edges, shadows, highlights) is predicted, given a scene hypothesis. Predictions and actual observations could then used in a Bayesian filter to update an estimate of the scene.

This is just a rough sketch of course. But maybe observing human performance in such visually challenging situations can point the way to technical solutions that degrade equally gracefully.

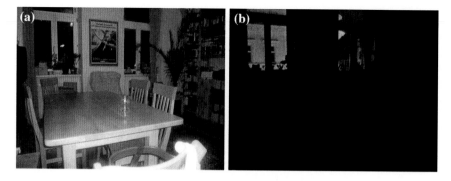

Fig. 9.1 Glass on a table in normal viewing conditions (**a**) and in very low light (**b**). Reconstructing the scene in the latter case or detecting the glass using a geometrical model is very hard, even for humans. Confirming a given hypothesis from a few reflections on the glass surface however is far easier and suffices to locate the glass (images best viewed in *colour*)

9.5 Conclusion

Sloman (1978) quite correctly predicted that by the end of the twentieth century (and it is just as true more than a decade later) computer vision would not have progressed enough to be adequate for the design of general purpose domestic robots, and that only specialised machines (with specialised abilities) would be available. This is indeed the state computer vision is in today.

The power of biological vision systems seems not to lie in perfect sensors and processing results, but in dealing with imperfect ones. Failures, uncertainties and ambiguities are not exceptional states of an otherwise perfectly functioning system, but instead part of the normal flow of processing.

We should thus aim at understanding the powerful computational principles allowing biological vision to infer a sufficiently accurate model of reality from partial, ambiguous, sometimes erroneous information derived from various cues. Actually this is already happening within many of the approaches presented above in the form of probabilistic models, however not on a system wide level.

Bringing the individual successful pieces of vision together into an equally successful system that eventually allows robots to operate within the challenging environments of our apartments, remains a ambitious goal.

Acknowledgments The research leading to these results has received funding from the European Community's Seventh Framework Programme FP7/2007-2013 under grant agreement No. 215181, CogX No. 600623, STRANDS the Austrian Science Foundation (FWF) under grant agreement No. I513-N23. vision@home No. TRP 139-N23, InSitu.

References

Agarwal S, Snavely N, Simon I, Seitz SM, Szeliski R (2009) Building Rome in a day. In: Proceedings of the international conference on computer vision, pp 72–79

Bay H, Ess A, Tuytelaars T, Van Gool L (2008) SURF: speeded up robust features. Comput Vis Image Underst 110(3):346–359

Biederman I (1987) Recognition-by-components: a theory of human image understanding. Psychol Rev 94(2):115–147

Biegelbauer G, Vincze M, Wohlkinger W (2010) Model-based 3D object detection: efficient approach using superquadrics. Mach Vis Appl 21:497–516

Binford TO (1971) Visual perception by computer. In: Proceedings of the IEEE conference on systems and control

Björkman M, Kragic D (2010) Active 3D scene segmentation and detection of unknown objects. In: 2010 IEEE international conference on robotics and automation, pp 3114–3120

Brooks R (1983) Model-based 3-D interpretations of 2-D images. IEEE Trans Pattern Anal Mach Intell 5(2):140–150

Chestnutt J, Kagami S, Nishiwaki K, Kuffner J, Kanade T (2007) GPU-accelerated real-time 3D tracking for humanoid locomotion. In: Proceedings of the IEEE/RSJ international conference on intelligent robots and systems

Choi C, Christensen HI (2010) Real-time 3D model-based tracking using edge and keypoint features for robotic manipulation. In: IEEE international conference on robotics and automation, pp 4048–4055

Clowes MB (1971) On seeing things. Artif Intell 2(1):79–116

Collet A, Berenson D, Srinivasa SS, Ferguson D (2009) Object recognition and full pose registration from a single image for robotic manipulation. In: Proceedings of the IEEE international conference on robotics and automation, pp 3534–3541

Cummins M, Newman P (2010) Appearance-only SLAM at large scale with FAB-MAP 2.0. Int J Robot Res 30(9):1100–1123

Dalal N, Triggs B (2005) Histograms of oriented gradients for human detection. In: IEEE conference on computer vision and pattern recognition, vol 2, pp 886–893

Davison AJ, Reid ID, Molton ND, Stasse O (2007) Monoslam: real-time single camera SLAM. IEEE Trans Pattern Anal Mach Intell 29(6):1052–1067

Dickinson S (2009) The evolution of object categorization and the challenge of image abstraction. In: Dickinson S, Leonardis A, Schiele B, Tarr M (eds) Object categorization: computer and human vision perspectives. Cambridge University Press, Cambridge, pp 1–37

Dickinson S, Pentland A, Rosenfeld A (1992) 3-D shape recovery using distributed aspect matching. In: IEEE Trans Pattern Anal Mach Intell 14(2):174–198

Fei-Fei L, Fergus R, Perona P (2006) One-shot learning of object categories. IEEE Trans Pattern Anal Mach Intell 28(4):594–611

Ferrari V, Tuytelaars T, Van Gool LJ (2006) Simultaneous object recognition and segmentation from single or multiple model views. Int J Comput Vis 67(2):159–188

Fitzgibbon AW, Zisserman A (2000) Multibody structure and motion: 3-D reconstruction of independently moving objects. In: Proceedings of the European conference on computer vision, Springer, pp 891–906

Fitzpatrick P, Metta G (2003) Grounding vision through experimental manipulation. Philos Trans Math Phys Eng Sci 361(1811):2165–2185

Gordon I, Lowe DG (2006) What and where: 3D object recognition with accurate pose. In: Ponce J, Hebert M, Schmid C, Zisserman A (eds) Toward category-level object recognition, Springer, Heidelberg, pp 67–82 (chap What and w)

Hager GD, Wegbreit B (2011) Scene parsing using a prior world model. Int J Robot Res (12):1477–1507

Hinterstoisser S, Holzer S, Cagniart C, Ilic S, Konolige K, Navab N, Lepetit V (2011) Multimodal templates for real-time detection of texture-less objects in heavily cluttered scenes. In: IEEE international conference on computer vision

Hoiem D, Efros A, Hebert M (2006) Putting objects in perspective. In: Proceedings of the IEEE conference on computer vision and pattern recognition, pp 2137–2144

Huffman D (1971) Impossible objects as nonsense sentences. Machine intelligence 6. Edinburgh University Press, Edinburgh

Intraub H (1981) Rapid conceptual identification of sequentially presented pictures. J Exp Psychol Hum Percept Perform 7:604–610

Klein G, Murray D (2006) Full-3D edge tacking with a particle filter. Proc Br Mac Vision Conf 3:1119–1128

Klein G, Murray D (2007) Parallel tracking and mapping for small AR workspaces. In: Proceedings of sixth IEEE and ACM international symposium on mixed and augmented reality (ISMAR), Nara, Japan, pp 225–234

Kraft D, Pugeault N, Baseski E, Popovic M, Kragic D, Kalkan S, Wörgötter F, Krüger N (2008) Birth of the object: detection of objectness and extraction of object shape through object action complexes. Int J Humanoid Rob 5(2):247–265

Leibe B, Schiele B (2003) Interleaved object categorization and segmentation. In: Proceedings of the British machine vision conference

Lepetit V, Fua P (2005) Monocular model-based 3D tracking of rigid objects: a survey. Found Trends Comput Graphics vision 1(1):1–89

Lourakis MIA, Argyros AA (2009) SBA: a software package for generic sparse bundle adjustment. ACM Trans Math Software 36(1):1–30

Lowe DG (1999) Object recognition from local scale-invariant features. In: Proceedings of the international conference on computer vision, pp 1150–1157

Lowe DG (1987) Three-dimensional object recognition from single two-dimensional images. Artif Intell 31(3):355–395

Lowe DG (2004) Distinctive image features from scale-invariant keypoints. Int J Comput Vision 60(2):91–110

Marr D, Nishihara H (1978) Representation and recognition of the spatial organization of three-dimensional shapes. Proc R Soc Lond B 200(1140):269–294

Marr D (1982) Vision: a computational investigation into the human representation and processing of visual information. W. H. Freeman, San Francisco

Matas J, Chum O, Martin U, Pajdla T (2002) Robust wide baseline stereo from maximally stable extremal regions. Proc Br Mach Vision Conf 1:384–393

Mikolajczyk K, Schmid C (2004) Scale and affine invariant interest point detectors. Int J Comput Vision 60(1):63–86

Mörwald T, Kopicki M, Stolkin R, Wyatt J, Zurek S, Zillich M, Vincze M (2011) Predicting the unobservable: visual 3D tracking with a probabilistic motion model. In: Proceedings of the IEEE international conference on robotics and automation, pp 1849–1855

Mörwald T, Prankl J, Richtsfeld A, Zillich M, Vincze M (2010) BLORT—the blocks world robotic vision toolbox. In: Best practice in 3D perception and modeling for mobile manipulation (in conjunction with ICRA 2010)

Murphy-Chutorian E, Trivedi MM (2008) Particle filtering with rendered models: a two pass approach to multi-object 3D tracking with the GPU. In: CVPR workshop on computer vision on GPU's (CVGPU), pp 1–8

Nevatia R, Binford TO (1977) Description and recognition of curved objects. Artif Intell 8:77–98

Newcombe RA, Davison AJ (2010) Live dense reconstruction with a single moving camera. In: IEEE conference on computer vision and pattern recognition, pp 1498–1505

Nistér D, Naroditsky O, Bergen J (2006) Visual odometry for ground vehicle applications. J Field Rob 23(1):3–20

Ozden KE, Schindler K, Gool LV (2010) Multibody structure-from-motion in practice. IEEE Trans Pattern Anal Mach Intell 32:1134–1141

Özuysal M, Lepetit V, Fleuret F, Fua P (2006) Feature harvesting for tracking-by-detection. Proc Eur Conf Comput Vision 3953:592–605

Özuysal M, Fua P, Lepetit V (2007) Fast keypoint recognition in ten lines of code. In: IEEE Conference on computer vision and pattern recognition, pp 1–8

Pan Q, Reitmayr G, Drummond T (2009) ProFORMA: probabilistic feature-based on-line rapid model acquisition. In: Proceedinge of the British machine vision conference, pp 1–11

Pilet J, Lepetit V, Fua P (2007) Fast non-rigid surface detection, registration and realistic augmentation. Int J Comput Vision 76(2):109–122

Richtsfeld A, Mörwald T, Prankl J, Zillich M, Vincze M (2012) Segmentation of unknown objects in indoor environments. In: Proceedings of the IEEE/RSJ international conference on intelligent robots and systems

Roberts LG (1965) Machine perception of three-dimensional solids. In: Tippett JT (ed) Optical and electro-optical information processing. MIT Press, Cambridge, pp 159–197

Rosten E, Drummond T (2006) Machine learning for high-speed corner detection. In: Prococeedings of the 9th European conference on computer vision, pp 430–434

Rusu RB, Blodow N, Marton ZC, Beetz M (2009) Close-range scene segmentation and reconstruction of 3D point cloud maps for mobile manipulation in human environments. In: Proceedsings of the IEEE/RSJ international conference on intelligent robots and systems, pp 1–6

Sánchez JR, Álvarez H, Borro D (2010) Towards real time 3D tracking and reconstruction on a GPU using Monte Carlo simulations. In: 9th IEEE international symposium on mixed and augmented reality (ISMAR), pp 185–192

Sloman A (1978) The computer revolution in philosophy: philosophy, science and models of mind. Harvester Press (and Humanities Press), Hassocks

Sloman A (1989) On designing a visual system: towards a gibsonian computational model of vision. J Exp Theoret AI 1:289–337

Snavely N, Seitz SM, Szeliski R (2006) Photo tourism: exploring photo collections in 3D. In: SIGGRAPH Conference Proceedings, pp 835–846

Thorpe SJ, Imbert M (1989) Biological constraints on connectionist modelling. In: Connectionism in Perspective. Elsevier, Amsterdam, pp 63–92

Thrun S, Burgard W, Fox D (2005) Probabilistic robotics. MIT Press, Cambridge

Tola E, Lepetit V, Fua P (2008) A fast local descriptor for dense matching. In: IEEE conference on computer vision and pattern recognition, pp 1–8

Ückermann A, Haschke R, Ritter H (2012) Real-time 3D segmentation of cluttered scenes for robot grasping. In: Proceedings of the IEEE/RSJ international conference on intelligent robots and systems

Waltz D (1975) Understanding line drawings of scenes with shadows. In: Winston PH (ed) The psychology of computer vision. McGraw-Hill, New York, pp 19–91

Welke K, Issac J, Schiebener D, Asfour T, Dillmann R (2010) Autonomous acquisition of visual multi-view object representations for object recognition on a humanoid robot. In: Proceedings of the IEEE international conference on robotics and automation, pp 2012–2019

Zach C, Pock T, Bischof H (2007) A duality based approach for realtime TV-L1 optical flow. Pattern Recogn 4713:214–223

Chapter 10
The Rocky Road from Hume to Kant: Correlations and Theories in Robots and Animals

Jeremy L. Wyatt

10.1 Introduction

In the Computer Revolution in Philosophy Sloman (1978) makes the point that there is no perception without prior knowledge and abilities. This was a point first made by Kant (2004) in response to Hume's work on causation Hume (2008). Simply put, the Kantian view supposes that in order to have the notion that one event is caused by another, an agent must have a pre-existing causal theory. Hume's view is that some events simply tend to be followed by certain other events, and that all we can know is what we induce from experience, with the causal relationship being essentially a correlational one.

An example that illustrates the difference is useful. In several talks, Sloman has given the example of two cogs that are meshed and will turn together. He claims that if an agent has a Humean model of causation all it knows is that when one cog turns, so does the other. If, on the other hand, the agent has what Sloman characterises as a Kantian model of causation including prior notions such as that of impenetrability, it can reason about why one cog turns when the other does, even without ever seeing the two cogs turn. Sloman speaks of the motion of one cog as being determined by the behaviour of the other and the rules of physics. In addition, the agent with a Kantian model can reason about what happens if we pull the cogs apart since their teeth will no longer have contacting surfaces. The Humean model cannot account for this except by learning the dependence of the correlation on a variable such as distance.

What matters in robots and animals is whether or not their predictions and explanations are good ones. In other words, do they serve the needs of the animal or robot? More specific criteria include whether they are accurate enough, timely enough, and able to generalise when needed. First, we will consider the requirements on predic-

J. L. Wyatt (✉)
School of Computer Science, University of Birmingham, Edgbaston, Birmingham
B15 2TT, UK
e-mail: j.l.wyatt@cs.bham.ac.uk

J. L. Wyatt et al. (eds.), *From Animals to Robots and Back: Reflections on Hard Problems in the Study of Cognition*, Cognitive Systems Monographs 22, DOI: 10.1007/978-3-319-06614-1_10, © Springer International Publishing Switzerland 2014

tion for robots and animals. Following that this essay will describe real robot systems that can predict what happens to objects when they are pushed. Some of the robot algorithms presented can be characterised as Humean in nature, some as Kantian. Having sketched in qualitative terms how these methods work this essay will then spend some time reflecting on how it might be possible for a machine to have both a Kantian style model and a Humean one. As part of this, we will think about levels of description in scientific theories; and the idea that it may be advantageous to have many micro-theories rather than relatively few macro-theories.

10.2 The Benefits of Prediction

What do animals and robots use predictions for? Prediction is concerned with stating what will happen over some future time period, given the current estimated state of the world and some control strategy on the part of the agent. The ability to predict what is going to happen next is one of the most useful abilities that animals can have. If you are good at predicting what happens next, you have several benefits. First, if the world changes rapidly you can use predictions to make decisions about what to do next without waiting for feedback. This is a kind of feed-forward control, and it works better than feedback control in situations where the feedback loop is quite slow relative to the pace of change in the world. Recently Webb (2004) has argued that some kinds of insects use prediction in just this way. Such strategies are not restricted to insects however. There is good evidence that systems such as the ocular-motor control and gripping systems in primates also use prediction to enable feed-forward control (Flanagan and Wing 1997; Kawato 1999), and that the cerebellum, which has a role in creating smooth generation of behaviour, also has the ability to predict (Miall and Wolpert 1996). In biology machines that predict are commonly referred to as forward models (or internal models).[1] Miall and Wolpert (1996) suggest that besides feedforward control, forward models can be used among other things for cancelling out the sensory effects of actions, supporting learning, performing state estimation, and mental planning and practice. We will now focus on that last purpose.

The idea of using a forward model for mental planning goes back at least to Kenneth (Craik 1943). Craik emphasised the utility of models of the environment, and of models of how the environmental state changes under actions of an agent. Craik put such models at the heart of his picture of cognition because only if an agent can predict the outcomes of actions in new situations can it plan to act in those situations. In familiar situations, a reasonable strategy is simply to learn a direct mapping from sensing to action. But in a new situation planning is required to obtain a new strategy. Planning will be particularly beneficial if the agent is in possession of forward models that generalise well to new situations. Generalisation in machine learning is typically used to refer to what is known as interpolation. Intuitively this is when the novel situation falls within the range of previous experience, even though it

---

[1] I will use the term forward model interchangeably with that of predicting machine in what follows.

is strictly different from all other previous experiences. Another kind of generalisation is extrapolation. Models that can extrapolate to situations that are new, but which in particular fall outside the range of previous experiences are applicable in a much wider range of circumstances, but also much harder to represent and acquire. There is also the issue of generalisation with respect to the actions to be executed. Forward models that predict the effects of simple or atomic action sequences are clearly more flexible in that their predictions can be easily sequenced to produce predicted outcomes for longer chains of novel behaviour. On the other hand it also makes sense that such sequenced predictions might well be less reliable than predictions tailored to specific, longer sequences of action.

Finally a forward model can also be used as a simulation of the world for learning control (mental rehearsal for learning). This strategy is widely used in AI style adaptive control, such as reinforcement learning. In this case, we can think of the model as a sample generator. The advantage of such an approach to learning is that the feedback controller once learned can be run reactively, and is thus quicker to use at run-time than invoking of a new planning process. Biological evidence for brain structures that are involved in planning and learning by running internal models is thinner on the ground, but it seems self-evident that both occur in a variety of animals, certainly such strategies have been successfully employed in many artificial systems.

10.3 Prediction in Robots

The previous section looked at why prediction might be useful. In this section, we will look at a series of realised artificial machines that perceive and make predictions about the behaviour of objects. For each type of machine, we can ask the question whether or not it fits better with one or other of the two characterisations Sloman distinguishes.

For the past few years, we have been working on the problem of predicting what an object will do when manipulated by a robot (Kopicki 2010; Kopicki et al. 2009, 2011). To keep the experimental set up simple, but still allow a rich range of object behaviours we have looked at pushing objects with a single robot finger rather than grasping with multiple fingers or in-hand manipulation. Pushing allows us to produce object behaviours that depend significantly on variables other than the pushing trajectory itself, such as frictional coefficients, the object shape, the pose, number and shape of surfaces the object is in contact with, and the mass distribution of the object. It is important to note that most of these are not directly unobservable. The object and the robot finger are both restricted to be rigid. Figures 10.1, 10.2 and 10.3 show three motion sequences of objects of different shape being pushed across a table. In the case of the L-shaped flap, which we refer to as a polyflap, it can twist, slide, tilt, and topple over or topple back to its original pose. The cylinder can roll, slide or twist, and the box can slide, twist and topple over. In many cases, it is possible for an object to do two or three of these at the same time. The physics of each situation is simple in theory, but in practice it is hard to predict exactly what each object will do, even

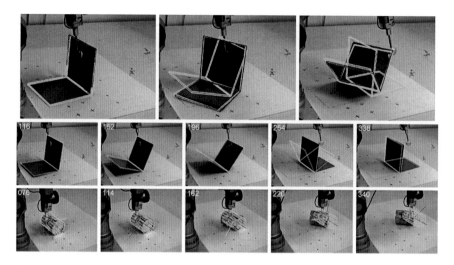

Fig. 10.1 Machine B predictions. (*Top row*) The *red wire-frame* shows the output from the vision tracking system. The *green wire-frame* indicates the object pose predicted by Machine D, while the *blue wire-frame* is Machine B. Although the predictions of Machine B are qualitatively plausible, it was virtually impossible to learn the physical parameters of the objects so that its predictions match reality in all cases. Note that the entire motion sequence is predicted before the physical push is initiated, without any correction from visual feedback during the push execution. (*Middle row*) Here the *green wire frame* shows Machine B's predictions for a box. (*Bottom row*) The *green wire frame* shows Machine B's predictions for a cylinder

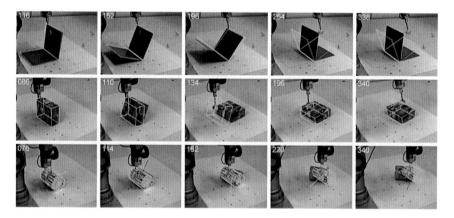

Fig. 10.2 *Green outline* shows predictions (from *top row* to *bottom row*) by Machine C on three different objects (The frame number is shown in the top left of each image)

given the full trajectory of the finger. Let us now describe three different machines that can produce predictions of how the object will move given the initial starting position of the object and the robot finger, and the planned trajectory of the robot finger. The goal of each predictor in this problem will be to produce a trajectory or a set of trajectories along which the object will move.

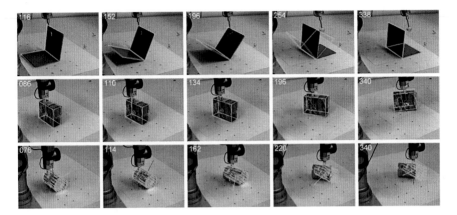

Fig. 10.3 The behaviours of three different objects when pushed. The *green wire frame* models represent predictions produced by a machine of type D as described later in the text (The frame number is shown in the *top left* of each image)

The predictors come in two different types, which for now I shall call Kantian machines (Machines A and B) and Humean machines (Machines C and D). Kantian machines have a priori knowledge built into them about rigid body physics, Humean machines do not and simply learn predictions that match or explain the data. All the machines produce predictions of what happens in the next time step only. To produce multi-step predictions the first prediction is fed back as a new input state along with the planned action for the second time step and the machine will then predict the state two steps into the future. This can be repeated to produce predictions arbitrarily far into the future. Now let us describe each machine in turn.

Machine A is a predictor that captures many of the notions of rigid body physics in a set of equations that it uses to calculate the motions of bodies under applied forces. There are many such predictors already in existence, called rigid body dynamics engines, and they are used for many different purposes. One advantage of such an approach to prediction is that a single physics engine is able to produce predicted trajectories for all possible rigid body interactions. Physics engines rely on the fact that Newton's laws are general in scope. This is both their power and their weakness. To produce a good prediction all of the typically unobservable properties of the objects mentioned above must be known. If we know the precise values of these properties then Machine A can produce a good prediction, but if it has to guess them then its prediction will most likely be wrong, as the laws governing rigid body dynamics are sensitive to small variations in their values.

Machine B is a variation on machine A. The difference is that machine B does not know a priori the specific parameters for the object or objects for which it is trying to make a prediction. It can, however, be shown many trajectories from a real object being pushed. We can then perform what is known as model selection (learning) of the parameters of the objects in the simulation. A simple way to do this is to guess a set of parameters (mass, friction, etc), and to run a simulation with those parameters and see how well the predicted behaviour matches the actual behaviour. We can keep generating and testing parameter sets until we find one that is sufficiently good. Thus, we have used the physics simulator to identify the underlying physical parameters that were otherwise unobservable using our knowledge of mechanics as captured in the physics simulator. Note that Machine B will now produce good predictions for the object it was trained on, but it will need to be trained again for each additional object.

Machine C is a correlational learner. Specifically, we set up the problem of learning to predict as a regression problem. This means treating the problem as one of finding a mapping from the current pose of the finger, the object, and the motion of the finger, to the motion of the object. As in the case of the physics engine, time must be quantised. As in the case of machine B, we train the machine by playing machine C many trajectories so that it learns the mapping. There are many machine learning techniques able to acquire such a mapping, and the challenges typically lie in learning a mapping that interpolates well. Typically, the problem gets harder as the number of input parameters increases. If we want to train Machine C to predict for different objects, we can either include a parameterisation of the object (its mass, frictional coefficients, etc) as inputs to the machine, or we can create a blank copy of machine C and train that copy on a new object. In the first case, machine C will have as many input dimensions as there are parameters that are needed to characterise all the objects. So again we need to know the typically unobservable properties of the objects. If at run time, we have a new object we can perform estimation over the space of parameters to find out which of several trained objects the new object is most like. In the case of creating a blank copy of machine C we effectively do the same thing, but without explicitly having to know anything about the object parameters. This means that each copy of Machine C learns to predict for a specific object, but it does not have an explicit parameterisation of the mass, frictional coefficient, etc that tells us exactly how each copy of Machine C differs from each other copy.

Machine D is also a correlational machine like Machine C, but instead of learning a mapping which makes a single prediction (regression) can produce multiple predictions. An important property of Machine D is that it produces predictions that have the same relative frequency as in the real world. This is important in two ways. First, it makes the assumption that many different outcomes are possible given the same state and action. The assumption is that the change of the world state cannot be determined because it is simply not possible to observe all the variables precisely enough to do so. In other words, it is not possible to observe the frictional coefficients, mass, etc, precisely enough that the object's behaviour is determined. In such a case, it is possible either to characterise the typical response and the typical degree of variation in the response, or to produce a full description of all different possible outcomes and their relative frequencies. In machine learning terms, this latter case boils down to posing the problem as a density estimation problem. In our scenario, this means the learning problem is to learn a probability density over the space of possible rigid body transformations an object can undergo given an initial pose and the applied trajectory of the finger.

The predictions made by some these machines are shown above. In fact we only show predictions for Machines B, C, and D, since machine A is either clearly inferior to machine B for any specific object, or requires far more knowledge than is directly available from the motion sequences. The results are shown in Figs. 10.1, 10.2 and 10.3. The predictions produced by each machine are broadly similar. Each machine (Kantian or Humean) must be trained separately for each object. Sometimes Machine B does not do as well as either of the Humean machines. For example, in the top row of Fig. 10.1 it is the case that Machine B fails to predict the rotation of the polyflap, and incorrectly predicts that it flips over. This is because its model inevitably makes some simplifications in order to be general, and even good training of its parameters cannot overcome this for all motion sequences for a particular object. Machine C gets stuck at the end of the top row in Fig. 10.2 whereas Machine D does not (Fig. 10.3 top row). None of the machines gives a perfect dynamic simulation, however, the predicted motion sometimes lagging the real motion.

We have now seen that it is possible to produce predictions in several ways. We saw that each predictor can be specialised to produce predictions for each different object. Importantly each predictor is also able to produce predictions in new situations that are similar to experienced situations. The predictions of the physics engine (Machine B) are sometimes inferior in accuracy to those of C and D, but they are applicable to a much wider range of shapes. The Humean machines, C and D, are able to specialize their predictions for each object, and thus produce high quality predictions, but their predictions are limited to those objects. We also saw that the density learning machine did a little better than the regression machine, but that there was not much difference in their performance.

The Humean machines do contain a little bit of prior knowledge. Each machine is built to monitor certain variables, so that it is known what is relevant to prediction. Second, the perceptual processing that visually tracks the object as it moves requires information about the shape of the object to be tracked, and this is true of all the machines. Third, to recover the pose of the object from a monocular image the machine needs to have some prior knowledge of geometry. Thus, although I characterised machines C and D as Humean, they still contain types of knowledge which provide a framework within which the correlational learning must occur.

A more interesting problem that our machines are able to solve is that of generalisation. Machine A generalises without further effort to new motion sequences, with new objects, both interpolatively and extrapolatively and with respect to state and action, if the parameters for the object in question are known. Machine B, which will of course consist of a set of machines, each with physics engine parameters tuned to a particular object, should be able to generalise extrapolatively with respect to shape, and interpolatively with respect to the physics engine parameters it has tuned. The interpolative generalisation can be handled by picking machines trained on objects of similar shape to the new object, and then clustering and averaging their predictions.

Machines C and D present a much more difficult problem with respect to generalisation. Interpolative generalisation is easy enough, since interpolation is the bread and butter of any modern machine learning algorithm. But extrapolative generalisation is much harder, since it is unreasonable to expect a correlational machine to

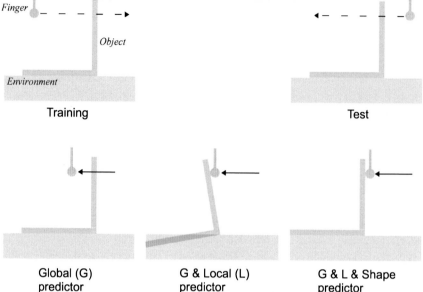

Fig. 10.4 Schematic diagram (2D projection of 3D scene) in which an object (of *L-shaped cross-section*) on a supporting surface is pushed by a robotic finger. Various predictors are trained solely on forward pushes (*top left*), but tested on backwards pushes (*top right*). The top panels show the push trajectory for the training and test phases, whereas the bottom panels show the outputs from three types of predictor in the test phase. A predictor comprised of just a global expert (Machine C or D) will fail to generalise, and will predict that the object does not move as the finger passes through it (*bottom left*). Adding a local expert will stop the finger penetrating the object, but does not guarantee that the predicted object motion will respect other impenetrability constraints (*bottom middle*). Finally, using an additional 'local shape' expert attached to the base of the object, a physically plausible motion is obtained (*bottom right*). Machine E is a machine with two or more experts

predict well on a shape or action that lies outside the range of its previous experience. One approach we tried with some success is to turn the extrapolation problem into an interpolation problem by using a particular representation. The intuition behind this approach is depicted in Fig. 10.4, and this is the intuition applied in Machine E.

The core idea is that instead of learning one predicting machine for each object, several machines are learned. Each machine is specialised in that it learns about the behaviour of one part of the object, ignoring non-local information. The particularly important parts of the object are those that are in contact with other surfaces, such as the robot's finger, or the surface of the table. In the simple case shown in the figure, three machines are learned for the one object, and each learns the constraints that a particular contact provides. Since the world of the robot is composed of rigid bodies these constraints are essentially impenetrability constraints. The prediction for the overall object requires the learner to combine the predictions of each of the

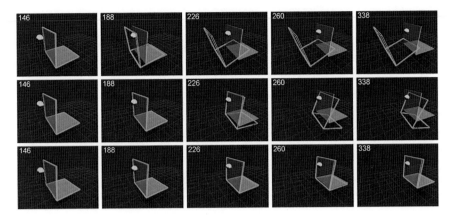

Fig. 10.5 Machine E: Extrapolative generalisation to Action. The green outline shows predictions (from *top row* to *bottom row*) by Machine D, a two expert version of Machine E, and a three expert version of Machine E, compared to simulated 'ground truth' (*in cyan*). These predictions illustrate exactly the rationale for multiple experts presented in Fig. 10.4. Machine D is a global predictor from Fig. 10.4, the two expert version of Machine E has a global and a local expert, and the three expert Machine E has a global, a local and a shape expert (The frame number is shown in the *top left* of each image)

predictors. We call these predictors experts. In a regression framework combining the outputs of the predictors is very difficult, since each machine produces only one predicted motion. In a density estimation framework this problem becomes much easier, since the predictors produce probability densities. A simple approach to combination then becomes to take a product of the densities, i.e., to estimate the likelihood of a particular motion as being the product of the likelihoods of that motion according to each expert. This is known as a product of experts machine, and is well known in the machine learning literature. In our problem, it enables powerful extrapolative generalisation with respect to shape and action, as shown in Figs. 10.5 and 10.6. In this figure, for example, the learners were trained using a small number of actions or shapes, and then the resulting predictors were presented with prediction problems involving actions or shapes outside of the range of those previously experienced. In the case of shape, the machines are trained on data from a cylinder and a polyflap, and then queried on the behaviour of two cylinders connected together. This is a very hard problem, and the experts currently have to be selected by hand and attached to the new object, so there is human intervention. This expert allocation process can be handled automatically, however, and this is a topic for future work.

What is important, however, is that we have seen that extrapolative generalisation is possible with correlational machines. In this sense, they are able to match, albeit in a more limited way, the generalisation power that physics engines possess. This is important since extrapolative generalisation is what intuitively makes Kantian machines—machines that really have theories of the world, rather than knowledge of correlations—so appealing. Turning back to the example of the cogs, it is perfectly

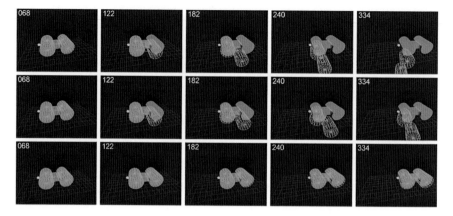

Fig. 10.6 Machine E: extrapolative generalisation to novel shape (simulation). The *green outline* shows predictions (from *top row* to *bottom row*) by Machine D, a two expert Machine E, and a three expert Machine E, compared to simulated 'ground truth' (*in cyan*). Note that the 1 and 2-expert methods predict that the object moves into and through the ground plane (The frame number is shown in the *top left* of each image)

reasonable to say that Machine E, properly configured could produce predictions of exactly the kind that are required to model the meshed teeth of the cogs, despite never having seen the cogs.

But this is the world of rigid bodies, and most animals have evolved to interact with non-rigid bodies most of the time. What sorts of predicting machines can we build in such circumstances? In Arriola-Rios and Wyatt (2011) we have started to explore machines of type B to model and make predictions of the deformations of deformable objects. In this case, the physics model is a simple spring-mass model, and an algorithm is used to search the parameter space for that model so as to match a set of data, again movies of deformations of objects made of a particular material. We have, however, discovered a further important problem with such models. Simply put this is that the more physics knowledge is built into a model, and the less is free to variation through learning, the more restricted is the class of objects for which that model can learn to predict. This may seem obvious, but it is important. Consider the physics interactions of jam or butter being spread onto a piece of bread. The bread might break if the butter is hard. There are lumps in the jam, some of which are crushed depending on the forces applied. Yet, spreading butter and jam on bread is a scenario where humans are able to generate some kind of prediction. The question thus arises what kinds of prediction humans are making in that case. There are no obvious answers, but what seems clear is that detailed physics models are probably not a feasible technology for artificial prediction in these more complex cases, and instead machines such as machines C–E above hold more promise. Having now, however, provided some evidence as to the state of machine learning applied to prediction in robotics we will move on to the problem of what it means to have a scientific theory, and to consider whether that produces a useful analogy for thinking about the kinds of predictive machines that we wish our robots to have.

10.4 Micro and Macro Theories

What sorts of machines are there in terms of the Kantian-Humean divide? Does it make sense to think about a rigid divide between machines that have prior knowledge about causation and those that don't? In engineering terms it seems to make most sense to think of a spectrum. At one end are machines such as Machine A that clearly have no beliefs based on experience and rely entirely on their prior knowledge. In the middle are machines such as machine B which have prior knowledge, but have elements that are tuned by experience. Toward the other end are machines such as C and D which largely rely on experience. Machine E is arguably more like machine B in that it relies on the prior notion that the experts must agree, and that conditioning based on contact relations is a good idea.

So instead of having a sharp division between what is and isn't a theory, it seems better to think about how *theory-like* something is. The ability to predict, to generalise in extrapolative ways rather than merely in interpolative ones, to provide an ability to perform abductive reasoning to find causes, and to say what is and is not possible are all aspects of theoryness. By this measure the machines described are theory-like, but in quite limited ways. The next aspect to consider in determining theoryness, is to consider the scope of a theory, by which is meant the range of circumstances to which a theory applies. Theories such as Newtonian mechanics are very broad in the range of circumstances to which they apply. How broad do animals and robots need their predictive theories to be?

When considering this question it is important to address the issue of the purposes to which a prediction will be put. Natural selection tells us that the need for prediction is driven in creatures by the evolutionary advantage of better control. Prediction has no survival value in and of itself. Another way of putting this is to ask what brains are for. In terms of natural selection brains are to give creatures selective advantage by giving them better control. Better here is necessarily loosely defined. It could be the ability to generate a faster response, or slower but more precise control so that they can perform tasks (e.g. prying a bug from a hole) that give them a niche they can prosper in. What is better depends on the niche that a creature inhabits. But ultimately, in survival terms, it only matters what a creature does, not why, and the brain processes themselves are servants of that action selection process. So when asking what kinds of theories creatures might have of the world, the answer is ones that are useful in selecting better actions than they would have otherwise.

Given this it seems likely that predictive machines in animals will each be quite narrow in scope. Why is this? The answer lies in (i) the selective advantage of predicting well for circumstances that animals are likely to encounter and have an impact on fitness and (ii) the need for evolution to find a series of solutions each of which has an incremental benefit. If we consider prediction, we have already seen that there is evidence that insect control mechanisms include predictors, but that these are typically used for feedforward control, not for planning. Suppose that other types of animals also first evolved predictors for such a purpose. It seems likely that different systems were able to reuse this prediction trick, adapting it for different control needs. Animals may well have contained predicting machines of a variety of kinds even early in evolutionary history. It may also be advantageous to evolve a *specific* type of predictor for each different control system because the rate at which the predictor should work, the timescale over which it should predict, and the rate at which the controller and the world state update are intricately linked in determining the performance of the overall control system.

Once evolution has discovered the utility of predictions combined with planning, however, the balance shifts from specific to general mechanisms. Planning performance is not so time sensitive to the prediction mechanism employed. Planning is as ubiquitous as the prediction mechanisms it employs to imagine future world states. A plausible story is thus that evolution discovered the trick of feeding a predictor's output back into itself, and discovered the ability to learn predictors. Once a machine that can learn to predict has been discovered, it can be applied to a wide range of circumstances. But each learned machine has to be fairly simple or it will take a fairly long time to learn. Thus it is advantageous for two reasons to simply create lots of copies of a machine that learns to predict and apply those copies to lots of different domains in which prediction plus planning has an advantage. This will give fast learning times. It will also gives evolution a smooth route to the animals being able to predict in a wide range of circumstances. This is because each additional copy of the learning machine gives an incremental benefit. For these reasons it seems plausible to think about animals as being likely to have many micro-theories, i.e. predicting machines that work for very small domains. Thus, an animal like Betty the Crow might have one predictor for stripping twigs of particular species of tree, another for a particular plastic material such as wire, another for the behaviour of grubs of a particular species she might try to extract from a rotten tree stump. Each of these predictors would have a strong learning element, as well as some prior bias or knowledge that constrained learning. It is useful to contrast this idea of micro-theories with the idea of macro-theories that explain very broad domains, such as Newtonian mechanics. Animals can develop macro-theories, humans are evidence of that, but this is note where the initial evolutionary benefit of theory formation lay. Better to have many micro-theories than a few macro-theories if you want to survive in the jungle.

10.5 Levels of Description in Science

In science, theories are typically judged on two criteria. The first is predictive accuracy, the second is parsimony. Judged by these measures Newton's laws constitute a remarkably successful theory of mechanics. They are valued because they are remarkably compact, and yet provide excellent predictions for objects operating on range of spatial scales. The theory of relativity and quantum physics, while far more complex, are also valued for their astonishing predictive power. Sloman (1978) has criticised predictive power and elegance as inadequate measures of scientific worth on their own and additionally emphasises the importance of scientific theories in saying what is and is not possible. He also distinguishes between what is possible and what is probable. Defining what is meant by knowledge of what is possible is rather hard. But for now recall that in some of the predicting machines we saw in Sect. 10.3 they could produce not one, but many predicted trajectories for an object when pushed. The most advanced machines were able to learn impenetrability rules, albeit statistically, and then apply these to restrict the range of probable outcomes, by making some trajectories very unlikely. But unlikely is not the same as impossible. So we cannot conclude that these machines are reasoning about what is possible and impossible in the way that Sloman describes. Therefore, one question is whether it is possible to learn a machine that can reason about impossibility and possibility in anything like that way.

To get at this issue, we need to consider the problem of levels of description in scientific theories. For now let us consider that there many areas of science in which there are theories that are more or less good at predicting the behaviour of systems that are the subject of their study. These include theories of animal learning, population biology and mechanisms of heredity. Many of these theories can be expressed at several levels of abstraction. It is also the case that in many parts of science there has been a belief in reductionism, the belief that in principle theories expressed at one level of abstraction ought best be translated into theories that operate at a greater of detail. I'm not a fan of that kind of reductionism, but it seems intuitive to talk about the ability to have multiple levels of description that interlock with the layer above and below. Dawkins (2006) calls this hierarchical reductionism. To illustrate let's take the example of brain science. If we consider the equations that describe neurons as spiking units they leave out many details of how those neurons work in terms of how depolarisation spreads across the surface of a neuron. More detailed models can capture this by modelling the spread using differential equations and taking account of the surface shape and the distances between the synapses while doing so. Still more detailed models can explicitly capture some of the biochemical mechanisms by which this depolarisation actually occurs at the channel level. For the sake of argument, let's restrict ourselves to consider models that we can think of as having a temporal component, in other words they are models that can, given the state of some system at one time, predict the future states of the system. Thus, when we refer to a model we are referring to model of the current state, together with a set of rules of how that system state evolves.

The key with hierarchical reductionism is that predictions of system state in a model at one level of detail can be explained in terms of the details of the model at the next (greater) level of detail. In addition, the properties that are modelled at one level of abstraction emerge from the model at the greater level of detail. The advantage of a more detailed model is that it can explain why the abstract model behaves as it does. To really have an explanation counterfactual reasoning must be possible in the more detailed model to provide explanations for the model one above it in the abstraction hierarchy. In other words that we can set the parameters of one model to be different to the way they are normatively, i.e. in the real world, in order to show what conditions are necessary to generate particular phenomena. For example, we can explore the parameter space of a particular model (e.g. the parameters that govern how ion channels in neurons work) to show what kinds of behaviour are possible or not possible in the model above (e.g. when spiking behaviour is possible, and when it is not as a function of the characteristics of the cell). There will be parts of the space for the more detailed model that cause different emergent behaviours in the more abstract model. In that sense, the detailed model explains the behaviour of the model one level above it in the abstraction hierarchy.

More abstract models also have advantages. Notably that they are typically simpler to analyse or run, while producing predictions that may be better for some purposes. They also benefit from typically needing less information to make a prediction. However, they will obviously make predictions that are more abstract than more detailed models. Whether this matters depends on the purpose to which the predictions must be put. Previously I argued that a prediction could be used by creature in three ways: feed-forward control, planning, and learning. There are undoubtedly others, but for the purposes of the argument, let us think about those ones here. To be even more restricted let us look at this issues of abstraction by turning back to the pushing object domain.

Take the object pushing scenario previously considered. For low-level control purposes, it is useful to have metric forward models of their behaviour. These are necessary to be able to decide where exactly the object needs to be pushed to achieve a particular motion. But some control does not need to reason at such a level of precision. Perhaps the task is to push the object (e.g. a box) into the corner of an arena, or up against one of its walls. In this case, there are many configurations of the object that satisfy the goal. Perhaps it doesn't matter if the object tips over, but it does matter that its sides are aligned and touching the wall(s), and yet it doesn't matter which precise section of the wall it is touching. Each goal constraint specifies a set of states. Because of this it may be easier to reason about how to achieve the abstract state by applying complex actions (e.g. pick up the block, push the block till it is touching the wall) and their effects on an abstract representation of the state. The question we will answer now is how such an abstract representation of state can be extracted from the precise metric representation of state that the predicting machines of Sect. 10.3 employ.

10.6 Levels of Description in Machines

In Sect. 10.3 we saw the results of machines that are able to predict how objects behave in detail when they are pushed. The learners are able to extract from their environment, correlations between what has happened and what can happen next. Each machine produced a trajectory through a continuous space, and some of the machines were able to produce probability densities over trajectories. The most advanced machine (E) used multiple experts to produce compromise predictions that enabled it to learn on two objects (e.g. a box and a cylinder), and then predict on another object with a previously unexperienced shape (a double cylinder). None

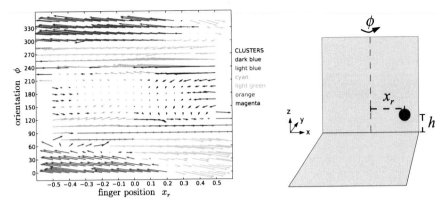

Fig. 10.7 (*Left panel*) Vector field, for $\delta = 0°$. Five major clusters were identified. (*Right panel*) Definition of push parameters h and x_r, and yaw angle ϕ. The tip of the robotic finger is depicted as a red sphere, which is just touching the upright plate of the polyflap. A *dashed line* marks the *vertical midline* of the polyflap

of the machines produce more abstract predictions. How could such a machine be obtained from these learned machines?

One answer to the issue of how to perform abstraction is how to look for what are called machine reductions, to find machines that are reductions of the predicting machines. To see an example of a machine reduction in our object pushing domain consider the fact that when pushing a box, it does not matter what absolute direction the box is pushed from in determining the outcome, but it does matter what direction the box is pushed from in the object's frame of reference. This means that equivalent pushes and orientations as we rotate the box (and the pushing finger around it by the same amount) together with the poses form a Lie group. Objects with symmetric shape will have further group structure. There are in fact algorithms for extracting Lie groups from continuous data as experienced by an agent, so it is not unreasonable to imagine an algorithm that would extract such groups from the predictions that the machines make. A simpler approach is to heuristically cluster predictions. In fact such a method can precisely be applied to the output of a physics engine, or a learned predictor. Each prediction sequence can be thought of as a vector sequence through space, and for a range of pushes a vector field can be defined. Figure 10.7 (left panel) shows a vector field for a range of pushes of an L-shaped polyflap extracted using just such a clustering algorithm. The right panel shows a typical parameterisation of a push, with the δ (not shown) giving the angle of the direction of push against the object.

The vector flow field is partitioned using a simple clustering algorithm. The clusters extracted are marked in different colours. To see the qualitative type of motion associated with a cluster, see Fig. 10.8 which shows a sample trajectory from the large light blue cluster. These very simple methods (as others) define state machine representations of the qualitative transitions of finger and pose, although uninformed by the contacts or force relations. Thus, this process of clustering gives a machine

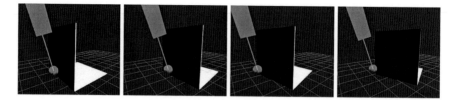

Fig. 10.8 Sample trajectory from the large *light-blue cluster* in Fig. 10.7. From *left* to *right*: the first two panels show the start and end configurations corresponding to the arrow at $(x_r, \phi) = (-0.05, 0)$. The last two panels show the start and end configurations corresponding to the arrow at $(-0.35, 15)$ (Note that the position of the polyflap is reset between the pairs of images (i.e. panel 2 and 3), creating an artefact in which the polyflap appears to move backwards)

reduction, and a move from a metric model of kinematics to qualitatively different motion changes.

What does the qualitative state representation achieve? It is still drawn from data, in that sense it is ultimately based on correlational models. Such a simple qualitative state representation could also not be used to constrain the predictions of a more detailed representation, i.e. to provide top down constraints. This would require a much richer representation. However, the idea of being able to find abstractions and then use these to influence or constrain prediction would be a step in the direction of the kind of models that Sloman advocates. There is a hard road ahead, however, to achieve the kind of models he terms "Kantian." Indeed, in order to find such abstractions powerful kinds of prior knowledge about symmetry, impenetrability, reversability, etc might be needed to begin with.

10.7 Summary

In this essay, we started with the difference between correlation and cause. Following this we focussed on the problem of how machines can learn predictive models of action outcomes in a continuous domain, and have seen that several practical methods are in fact correlational. Yet we have also seen that some correlational methods, when framed in the right kind of learning structure, are able to perform some quite powerful kinds of generalisation that are normally the preserve of models with far more prior knowledge (such as physics engines). I characterised the two types of machines as Humean and Kantian respectively. Following this, I have argued that including too much prior physics knowledge in such models is a bad idea, because the range of phenomena for which they predict is then rather narrow. This has then led to the idea that many models with relatively little knowledge are appealing for two reasons. First, they are amenable to learning and reallocation to new prediction tasks, and second they provide a mechanism by which we can believe that evolution discovered a royal road to machines that are capable of prediction for a wide range of phenomema.

The next part of the essay was rather more speculative. Essentially the argument went as follows. In science, there are many examples of theories that are linked, but which work at different levels of abstraction. We argued that this is a good idea in machines too, since it is efficient to predict and plan at a level of abstraction appropriate to the task, and it may often be necessary to plan at several levels of abstraction simultaneously. This in turn requires machines that can predict at several levels of abstraction. Finally, I showed one kind of approach to extracting abstract machines from the very low-level predicting machines described earlier in the paper. Nevertheless, the conclusion must be that finding such abstractions is a hard task.

Acknowledgments While the opinions expressed here are purely my own, most of the technical work has been carried out by Marek Kopicki, Sebastian Zurek, Rustam Stolkin, Thomas Mörwald and Vero Arriola-Rios. Thanks to them all. Many of the ideas for this work came from Aaron Sloman, including the practical idea of the polyflap domain. Thanks to Aaron for many hours of discussion, and for his kindness and support in my career.

References

Arriola-Rios VE, Wyatt JL (2011) 2D mass-spring-like model for prediction of a sponge's behaviour upon robotic interaction. In: Research and development in intelligent systems XXVIII, pp 195–208

Craik J (1943) The nature of explanation. Cambridge University Press, Cambridge

Dawkins R (2006) The blind watchmaker. Penguin, Hammondsworth

Flanagan JR, Wing AM (1997) The role of internal models in motion planning and control: evidence from grip force adjustments during movements of hand-held loads. J Neurosci 17(4):1519–1528

Hume D (2008) An enquiry concerning human understanding. Oxford University Press, Oxford

Kant I (2004) Prolegomena to any future metaphysics. Cambridge University Press, Cambridge

Kawato M (1999) Internal models for motor control and trajectory planning. Curr Opin Neurobiol 9(6):718–727

Kopicki M (2010) Prediction learning in robotic manipulation. Ph.D. thesis, University of Birmingham, Birmingham

Kopicki M, Wyatt J, Stolkin R (2009) Prediction learning in robotic pushing manipulation. In: International conference on advanced robotics, 2009. ICAR 2009, pp 1–6

Kopicki M, Zurek S, Stolkin R, Mörwald T, Wyatt J (2011) Learning to predict how rigid objects behave under simple manipulation. In: Proceedings of the IEEE international conference on robotics and automation (ICRA11). http://www.cs.bham.ac.uk/msk/pub/icra2011.pdf

Miall R, Wolpert D (1996) Forward models for physiological motor control. Neural Netw 9(8):1265–1279 (four Major Hypotheses in Neuroscience)

Sloman A (1978) The computer revolution in philosophy. Harvester Press, Hassocks. http://www.cs.bham.ac.uk/research/projects/cogaff/crp/

Webb B (2004) Neural mechanisms for prediction: do insects have forward models? Trends Neurosci 27(5):278–282

Chapter 11
Combining Planning and Action, Lessons from Robots and the Natural World

Jeremy Baxter

11.1 Introduction

In my work, I have been primarily concerned with getting robotic systems to act in real world environments. In the real world (or even realistic simulations), robots have to deal with a wide range of problems. They can have only a limited quantity of information about the world around them and a limited knowledge about what effects their own actions will have on that world. They have to interact with the world and the other systems and people within it, some of whom may be hostile, others may be co-operative. Finally, they have to understand what they wish to achieve and set about trying to achieve it.

Given the incredible layers of complexity presented by the real world, it can sometimes seem amazing that anything can be achieved at all. However, the world is full of animals and people who seem to manage the task very well! Given that the natural world represents a successful model taking as many lessons from it as possible can be important. In my work with robotic systems, I have always been happy to try to draw inspiration from the way animals and people have solved the problems of planning and acting in the real world.

At the first robotics and control conference, I attended Professor Ian Brady gave a speech in which he compared the then state of the art in robotics to the abilities of a sheep and pointed out all the ways in which sheep are superior. Sheep can detect the environment around them and move within it relatively freely. They can detect where they can or cannot go and move to accomplish their goals. They are not continually bumping into things or getting stuck in corners. They can interact with other sheep to achieve their own goals (and if we consider it at that level the 'goals' of the flock). They can also do things few robots can manage, continually re-fueling themselves and ultimately creating new sheep. All this out in the real world full of wind, rain,

J. Baxter (✉)
Poynting Institute, University of Birmingham, Birmingham, UK
e-mail: J.Baxter@Poynting.bham.ac.uk

J. L. Wyatt et al. (eds.), *From Animals to Robots and Back: Reflections on Hard Problems in the Study of Cognition*, Cognitive Systems Monographs 22, DOI: 10.1007/978-3-319-06614-1_11, © Springer International Publishing Switzerland 2014

mud, rocks and all sorts of hazards which would stump a robotic system and what is more they carry it out continuously, for years and years, while robots measure their run times in hours or minutes. We were, he concluded, a long way from producing robotic systems as competent in the real world as a sheep and despite the undoubted strides that have been taken in the intervening decades robotic systems still fall far short of the sheep level of performance.

Finding out how sheep, or indeed any animal or person manages themselves is a difficult problem in its own right, as is attempting to use this information to produce behaviour in a robotic system. It was working with Aaron Sloman however that introduced me to the concept that lessons could be be drawn back the other way. Aaron was determined to use computer models of decision-making, reasoning, and action to learn about how people acted and thought. By looking at the effect that different architectural components had on a computational reasoning system, we could gain a greater understanding into the reason that our brains work in the way that they do. I have only been able to make use of small steps and general principles in using natural systems to guide the design of artificial ones and using the knowledge gained from the performance of artificial systems to understand natural ones better. However, I have always been grateful that Aaron opened my eyes to a much wider field and greater goals than 'simply' engineering better robotic systems.

This paper considers the construction of architectures for combining planning and action for individual (robotic) agents and to a lesser extent for allowing co-operation on joint tasks by groups of such agents (multi-agent systems). I will describe the background to the types of problems these architectures are designed to deal with drawn from the research literature and relate how, in my opinion, they draw influence from and can inform the understanding of behaviour in people and animals. These sections are not intended to be comprehensive literature reviews but rather a set of personal influences related back to the wider goal, espoused by Aaron Sloman, of understanding ourselves better by understanding computational problems.

11.2 Planning and Execution Architectures

11.2.1 Introduction

Planning and acting are often divided into separate research fields. Planning attempts to identify the optimal set of action to achieve a goal (set of goals or some of a subset of goals (Briel et al. 2004)) with the assumption that this set of actions can then be executed perfectly. Control theory, on the other hand, assumes that the desired action is known but that feedback is needed to achieve it. Planners work on a domain model which gives them an internally manipulatable model of the world which can be used to consider the outcomes (or probabilistic range of outcomes) of a sequence of actions. Control systems, on the other hand, work with a directly sensed connection to the world and act to remove the detected error between the desired and the sensed states. Planners can take a considerable time to produce a plan and assume that the world is not changing while they are planning, control systems have to react fast to changing inputs.

Plan execution systems occupy the middle ground between these two types of decision system. They try to deal with situations where there is not time to gather all the the necessary information, update a full world model and run a planner before an action needs to be taken but the change is large enough that a feedback controller cannot compensate for it. Plan execution systems therefore have to balance time spent planning with time spent acting. More generally they need to balance the potential gains from additional planning to improve performance with gains (or losses) from executing or continuing to execute a sub-optimal plan.

A plan execution system can focus just on simple (often implicit) alternative ways of achieving steps in a plan but more powerful systems can also consider when and how to initiate planning, treating planners as resources available to the system. A general plan execution system therefore needs to consider the progress not only of actions it is trying to execute but any currently running planners. Changes in the world may make actions inappropriate or indicate they have failed. Similarly planning may become irrelevant or fail to identify a plan. Plan execution systems aim to manage the use of both execution and planning resources to give an improved overall outcome, typically this means reasoning at multiple levels of abstraction and scheduling both internal (computational) and external resources.

11.2.2 Exemplar Systems

There have been many examples of plan execution systems. An initial influential approach was Reaction Action Packages (Firby 1989). This was concerned with the generation of flexible plans to deal with cases where actions fail to produce desired effects, and when unexpected events demanded that the robot shift its attention. Rather than constructing a list of primitive robot actions to be executed a reaction action package (RAP) adds structure to the actions to allow *situation-driven execution*. Situation-driven execution assumes that a plan consists of tasks with three major components: a satisfaction test, a window of activity, and a set of execution methods that are appropriate in different circumstances. Execution of such a plan proceeds by selecting an unsatisfied task and choosing a method to achieve it based on the current world state. Effectively the output of a planner is augmented by descriptions of alternative actions that can be taken to achieve the same effect depending on the actual circumstances at the time of execution. This richer level of description allows the repair of executing plans even if the initial action fails. For example, an initial attempt to pick up an object may fail but be repaired by repeating the attempt with an alternative manipulator arm.

While RAPs envisaged that a planner would generate a sequence of action packages the Procedural Reasoning System (PRS) (Georgeff and Ingrand 1989; Lee et al. 1994) was intended to provide a means of encoding expert knowledge about alternative actions to provide robust execution. PRS is a general purpose reasoning system that is particularly suited for use in domains in which there are predetermined procedures for handling the situations that might arise (for example from previous experience or offline planning). PRS represents an instantiation of a Belief Desire Intention (BDI) architecture (Rao and Georgeff 1995). Rather than performing detailed planning in a changing environment Procedural Reasoning is based on matching current beliefs to knowledge about the appropriate plans (procedures) to carry out in that situation and executing them. The procedures can be stopped and started in response to both environmental changes and goal changes.

Several plan execution and diagnostic/fault recovery systems were developed by NASA. In particular, the system used on the Deep space 1 experimental mission (Gat 2009; Pell et al. 1997). The execution of plans on a spacecraft can be very challenging. Virtually, all resources on spacecraft are limited and carefully budgeted so planning and execution control must ensure that they are allocated effectively to goal-achievement. Some resources, like solar panel-generated power, are renewable but limited. Others, such as total propellant, are finite and must be budgeted across the entire mission. The planner used in Deep space 1 reasoned about resource usage in generating plans, but because of run-time uncertainty the resource constraints had to be enforced as part of execution. The planner and plan execution system reasoned about and interacted with external agents and processes, such as the on-board navigation system and the attitude controller. These external agents provided some information at planning time and could achieve tasks and provide more information at run-time but were never fully controllable or predictable. For example, the attitude controller provided estimates of turn durations at plan time, but the completion of turns during execution was not controllable and could only be observed and the consequent impact on a plan allowed for. The plans used on the spacecraft had to express compatibilities among activities, and the plan execution system synchronised these activities at run-time. In addition, planning and the information necessary to generate plans are also limited resources. Because of the limited on-board processing capabilities on spacecraft, the planner must share the CPU with other critical computation tasks such as the execution engine, the real-time control loops and the fault detection, isolation and recovery system. While planning is being carried out, the spacecraft has to continue to operate and some plans may contain critical tasks whose execution cannot be interrupted in order to install newly generated plans. Thus, the system scheduled planning sparingly and made use of embedded contingencies to ensure that execution was robust in the face of a wide variety of hardware faults and delays. This is an example of a system in which full re-planning was expensive and so detailed diagnostics and recovery procedures were necessary to ensure the set of high level planned tasks could be carried out.

The co-operative intelligent real-time control architecture (Circa) (Musliner et al. 1993) took this approach a step further by attempting to maintain hard real-time guarantees while still running complex reasoning where necessary. It consisted of separate AI and real-time subsystems, each addressing the problems for which they were designed. A structured interface allowed the subsystems to communicate without compromising their respective performance goals. CIRCA allowed reasoning about its own bounded reactivity and so could guarantee that it would meet hard deadlines while still using unpredictable AI methods. By using built-in knowledge about its abilities CIRCA attempted to trade off the timeliness, precision, confidence, and completeness of its output to meet the needs of a real-time control system.

11.2.3 Applications

In addition to the applications described above, I have been involved with additional uses of plan-execution systems in specific applications. As part of the 'Broad Agents' work carried out with Aaron's assistance I used the SimAgent toolkit to develop intelligent vehicles for use as Computer Generated Forces (CGF) in training simulations (Baxter and Hepplewhite 1999). This was then developed into a multi-agent system where the same plan-execution frameworks were extended to support the planning and co-ordination of actions for multiple vehicles (Baxter and Horn 2001). In many ways, this was a good environment for multi-agent plan execution systems since the military command structures it was modeling are very hierarchical and rely on pre-formed teams and authority structures enabling us to concentrate of developing and carrying out group plans rather than on the identification of potential teams and negotiation. While this was a good application environment for developing techniques, the level of sophistication was generally too high for the required applications. In most training scenarios, a small set of finite state machine-driven behaviours supervised by a human operator proved sufficient, and more flexible. The same techniques, however, had potential application in unmanned air vehicles (UAVs). In UAV applications, the operator is typically a long distance from the vehicle(s). Operators also have to devote considerable effort to managing and interpreting returns from sensor systems in addition to controlling the vehicles and their sensors. For these applications, we were able to use a multi-agent hierarchical planning and execution framework to control a team of vehicles carrying out collaborative tasks and using a range of different planning systems ranging from complex probabilistic search planners to simple reactive behaviours all in real time (Baxter et al. 2008). This work included test flights in real aircraft, in one case the pilot of one jet additionally controlling another jet acting as a surrogate unmanned aerial vehicle (with additional simulated teammates). This work showed the strength of planning and execution systems in the real-world but has left open the issues of the development time/cost and clearance into safety critical domains such as unmanned flight. Another application area is in autonomous driving, as shown in the DARPA urban challenge which had unmanned cars driving autonomously in small town. One entry in particular, from Cornell (Miller et al. 2008), used a three-layer planning and execution framework with behavioural, tactical and operational layers. The layers not only planned at different levels of abstraction but fed back progress and sensed information to higher levels to allow for re-planning if necessary.

11.2.4 Additional Approaches to Planning and Acting

One of the difficulties of applying plan execution approaches is that except in very narrowly defined problems they are inherently part of a much wider system that must be interacted with. The problems plan execution systems were designed to deal with may be addressed by other parts of a system if the environment is suitable. Plan execution is also dependent on the performance of both the planners and the control system it is interacting with meaning that designing the appropriate balance can be very difficult to achieve (and when dealing with legacy systems a complex compromise). If a planning system is suitably fast and accurate, it can negate the need for explicit plan execution. By continually re-planning changes in the environment can be dealt with as they appear (Yoon et al. 2007) and so the need to explicitly represent the actions to be taken when the situation changes is removed. At the other extreme, Markov decision processes (Russell and Norvig 2002) attempt to enumerate all actions to be taken in all possible states so that once planning has been completed in an uncertainty model execution becomes a simple lookup process. The rise in computer performance coupled with increasingly efficient planning methods has meant that domains in which it was previously impractical to run a planning system during execution are now able to make use of much more frequent re-planning to displace the need for a complex intermediate step between the planner and the control system. Conversely, a suitably complex controller using techniques such as model reference adaptive control (Landau 1979) can include complex forward models of a system giving better informed feedback control which may take into account some of the same factors that a plan execution architecture is designed to deal with. However, even though the boundaries may be moving there are still a wide range of situations in which explicit consideration of how to put a plan into action and when to generate, or re-generate new plans is necessary. In particular, long running systems which have to interact in real time with the world still face the challenge of balancing deliberation with action.

11.3 Cognition and Affect

Through working with Aaron I was introduced to the the Cognition and Affect project (Sloman 2011). At the time the main focus of my work was looking at agents to control vehicles within training simulations (Baxter and Hepplewhite 1999) and we were able to make use of the POPLOG toolkit that was developed at Birmingham to successfully produce a range of vehicle behaviours. One of the interesting features of the project was the focus on an architecture schema rather than a fixed decision-making system. The schema (see Fig. 11.1) incorporated three layers; Reactive, deliberative and meta-management. Information is filtered between these layers, usually moving up, and control flows down. The schema and the toolkit gave a lot of freedom to design different system architectures (probably too much freedom for

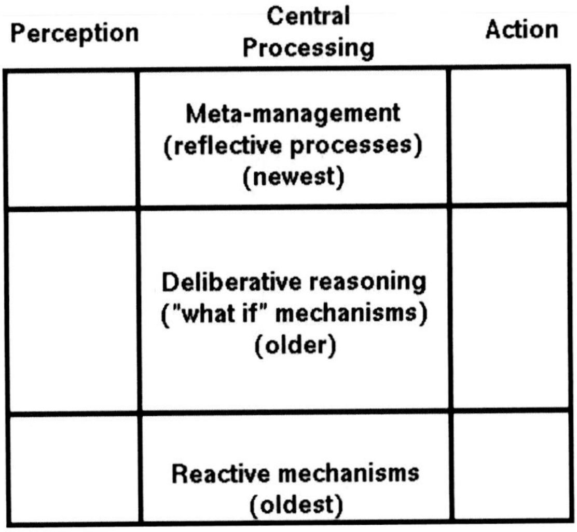

Fig. 11.1 The CogAff schema

Perception	Central Processing	Action
	Meta-management (reflective processes) (newest)	
	Deliberative reasoning ("what if" mechanisms) (older)	
	Reactive mechanisms (oldest)	

most practical applications!) and seeing the different approaches gave useful design pointers. In particular, what was key for our application was not providing a set of optimal plans in specific situations but to avoid looking stupid in as wide a range of situations as possible. For example, the toolkit was used to produce a model of grief (Wright et al. 1996) and while we had no need for full blown emotion models in our system it provided useful pointers for the features we needed to prevent our models becoming stuck in repetitive planning loops by recognising that emotion-like meta-management functions could moderate the deliberative and reactive layers in response to difficulties in both planning and reacting. This enabled us to avoid both long delays waiting for planners and too many repeat cycles in reactions (like a fly repeatedly hitting itself against a window). It is a long way, in my opinion, to claim vehicles were exhibiting frustration or fear but we could see how the architectural mechanisms could be linked.

Another powerful analogy we were able to draw on was in trying to prevent over planning. When developing motion behaviours in a complex and an uncertain environment, there is a tendency to repeatedly plan and to keep looking for the 'best' or 'optimal' answer. However, this has a tendency to cause vehicles to be indecisive and to spend long periods of time planning before they start moving which in some situations is worse than committing and acting. We had problems with a 'hide' behaviour for our vehicles, planning for which involved heavy and time-consuming line of sight calculations. Computing possible concealment was expensive and produced many different candidates. Committing too early to a location could lead to selecting a really bad hiding position, but waiting too long increased the probability that the vehicle would be unable to find and reach a hiding location in time. The selection mechanism we used was based on models of 'mate choice' (Real 1990). The core idea is to spend a certain period of time investigating the quality of different solutions, identifying the best so far and then to commit to the next solution which is found exceeding this threshold - or to revert to the previous best solution if no better solution is found in a fixed time. This gives a limit on the time taken to plan rather than act but also allows some time to characterise the quality of solutions which are available. This selection algorithm is drawn from models of mate choice where a 'choosy' female first examines potential mates to estimate the local distribution of mate quality and then adopts a threshold for choosing a mate based on this estimate.

11.4 Future Planning and Execution Architectures

While planning and execution architectures have had successful applications they are far from ubiquitous. There are a number of reasons for this that are often different between research and industrial applications. In many cases, a dedicated and an identifiable plan execution system is not necessary. If the domain is such that a planning system or controller can fully solve the system in the time available before execution is necessary then all that is needed is a regular re-planing cycle to notice and react to changes. In other cases, the available options are so small that the 'plan execution' system can be implemented as a small set of rules and, while necessary, is not an architectural component in its own right. Any planning system that interacts with the real world needs a translation layer to convert sensor information into an internal domain description and output actions into executable commands. If this translation layer also embeds some rules about how to plan and some execution options it is hard to distinguish it as a separate layer even though it is present and necessary. The point at which the re-use of the same planning and execution constructs mean that a separate framework is desirable can be hard to identify and so such systems often grow ad-hoc during development. If a plan execution system is regarded as a form of middleware, it has to make the same case as other middleware systems for inclusion. The complexity and development time in using and maintaining the system as well as translating to and from its own representation has to be balanced by the time it saves by reliably performing its function in multiple different parts of the overall system. We have certainly found that in a fast-moving research environment having a plan execution framework helps to develop and trial multiple different candidate solutions to a problem but the end result may be a system that does not need the addition of a dedicated plan execution layer. Sometimes, for example, a mechanism to drop into a safe fall back state and wait for further input is all that is required.

Safety clearance is also a critical issue for real-world systems. Often the execution of a system has properties which have to be adhered to maintain safety. The additional complexity of a dedicated plan execution system can make this harder to do. Conversely, if a plan execution layer can be shown to be safe and reliable, it can provide a powerful way of segmenting a safety case into smaller, easier to manage, elements.

To enable the current state of the art to be exploited, it will be necessary to generate:

- automated testing systems linked to safety cases for plan execution frameworks;
- design and prediction tools allowing the potential benefits to be identified;
- links to planning domain languages and control systems to make integration easier.

Future areas of research are I believe are:

- finding ways to predict and articulate the benefit of abstraction and multi-layer reasoning over a complete system;
- developing description languages for planning and execution components;
- developing techniques for gathering and distributing performance data about sub-systems
- learning to adapt and self-configure complex decision making systems as more data about the environment and the system's performance becomes available.

All of these research areas have direct analogues in understanding how people and animals react to their environment, and I believe that there will continue to be a valuable transfer of ideas between robotics and the study of biological behaviours.

References

Baxter J, Hepplewhite R (1999) Agents in tank battle simulations. Commun ACM 42:74–75

Baxter JW, Horn GS (2001) Executing group tasks despite losses and failures. In: Proceedings of the 10th conference on computer generated forces and behavioral, representation, pp 205–214

Baxter JW, Horn GS, Leivers DP (2008) Fly-by-agent: controlling a pool of uavs via a multi-agent system. Know-Based Syst 21(3):232–237. http://dx.doi.org/10.1016/j.knosys.2007.11.005

Briel MVD, Sanchez R, Do MB, Kambhampati S (2004) Effective approaches for partial satisfaction (over-subscription) planning. AAAI Press, San Jose, pp 562–569

Firby RJ (1989) Adaptive execution in complex dynamic worlds. Yale University (Tech. rep.)

Gat E (2009) Non-linear sequencing. IEEE Aerosp Electron Syst Mag 24(3):41–46. doi:10.1109/MAES.2009.4811088

Georgeff MP, Ingrand FF (1989) Decision-making in an embedded reasoning system. In: Proceedings of the Eleventh international joint conference on artificial intelligence. Morgan Kaufmann, Detroit (Michigan), pp 972–978

Landau YD (1979) Adaptive control: the model reference approach. Marcel Dekker Inc., New York

Lee J, Huber MJ, Durfee EH, Kenny PG (1994) Um-prs: an implementation of the procedural reasoning system for multirobot applications. In: Conference on intelligent robotics in field, factory, service, and space (CIRFFSS, pp 842–849

Miller I, Campbell M, Huttenlocher D (2008) Team cornell's skynet: Robust perception and planning in an urban environment. J Field Robot 25:493–527

Musliner DJ, Durfee EH, Shin KG (1993) Circa: a cooperative intelligent real-time control architecture. IEEE Trans Syst Man Cybern 23:1561–1574

Pell B, Gat E, Keesing R, Muscettola N, Smith B (1997) Robust periodic planning and execution for autonomous spacecraft. In: Proceedings of IJCAI-97, IJCAI. Morgan Kaufman Publishers, pp 1234–1239

Rao AS, Georgeff MP (1995) Bdi agents: from theory to practice. In: Proceedings of the first international conference on multi-agent systems (ICMAS-95), pp 312–319

Real L (1990) Search theory and mate choice. i. models of single-sex discrimination. Am Nat 136:376404

Russell SJ, Norvig P (2002) Artificial intelligence: a modern approach, 2nd edn. Prentice Hall, Englewood Cliffs

Sloman A (2011) http://www.cs.bham.ac.uk/axs/cogaff.html

Wright IP, Sloman A, Beaudoin LP (1996) Towards a design-based analysis of emotional episodes. Philos Psychiatry Psychol 3(2):101–126

Yoon S, Fern R, Givan A (2007) Ff-replan: a baseline for probabilistic planning. In: Proceedings of the international conference on automated planning and scheduling, pp 352259

Chapter 12
Developing Expertise with Objective Knowledge: Motive Generators and Productive Practice

Luc P. Beaudoin

12.1 Preface

Flush with scholarships and graduate school opportunities in 1990, having researched the Commonwealth for the most fertile ground in cognitive science, I heeded Dr. Claude Lamontagne's advice to study with a brilliant scholar he had known at the School of Artificial Intelligence of the University of Edinburgh (1972–1973), Prof. Aaron Sloman. Lamontagne praised Sloman's penetrating mind, one that always offered insightful comments, criticisms and suggestions aimed at the heart of the matter. Lamontagne also knew that Sloman (and Sussex University) fully embraced theoretical, computational cognitive science. He was right. Sloman is—as all who know him well will attest—a productive thinker of the highest caliber.

This chapter weaves five themes from cognitive science in my quest to understand and help enhance experts' cognitive productivity:

1. Productive learning and expertise (Bereiter and Scardamalia 1993; Wertheimer 1959). Sloman introduced me to the work of Wertheimer (e.g., Sloman 1978), who seemed to capture the essence of productive thinking, though, ironically, Wertheimer's understanding, like Freud's, was stunted by concepts from physics.
2. Motive processing, from a designer stance (Sloman 1993a). The stance itself affords productive thinking.
3. Conceptual analysis (Sloman 1978), which also are helpful thinking tools.
4. Potent psychological principles for productive learning (too many to list here).
5. Self-regulated learning with technology (Beaudoin and Winne 2009; Winne 2006).

This is a slightly updated version of the paper I presented at Aaron Sloman's festschrift (Beaudoin 2011). Many of the ideas presented here are developed further in Beaudoin (2013a, b).

L. P. Beaudoin (✉)
Simon Fraser University and CogZest, Burnaby, BC, Canada
e-mail: lucb@cogzest.com

The chapter makes frequent reference to Sloman's work, which is both an indication of how he has shaped my thinking and an illustration of the variety of ways in which his ideas can be applied and extended.

In my quest, I seek to wrest the mechanisms underlying "testing effects" (Kuo and Hirshman 1996) from Ebbinghaus's undying grasp on cognitive psychology, to polish them and to hand them over for theoretical and practical study to the scions of Immanuel Kant, Max Whertheimer, Frederic Bartlett and Warren McCulloch. I hope this chapter will inspire others to address the challenges I have posed.

12.2 Introduction

[We] must, so far as we can, make ourselves immortal, and strain every nerve to live in accordance with the best thing in us; for even if it be small in bulk, much more does it in power and worth surpass everything. [...] the best and the most pleasant life is the life of the intellect, since the intellect is in the fullest sense the man. So this life will also be the happiest.

Aristotle (Nic. Ethics, Bk. X, Chap. 7)

This chapter addresses factual and practical problems concerning expert multipurpose (broadly transferred) learning from objective knowledge (whether factual, practical, or normative (Sloman 1978, Chap. 2)). I describe purposes of and modern challenges to such productive learning. The factual problems concern the motive processing involved in productive learning. The practical problems are to enhance learning from knowledge resources. I focus on motive generators for the former problems. Based on some of the most potent applicable principles from cognitive science,[1] I propose productive practice for the latter. I raise a number of research challenges regarding these overlapping concerns.

I propose that when and after experts process knowledge resources[2] conveying objective knowledge in such a way that they can apply it much later—e.g., they develop a *lasting* understanding, acquire skills, develop new attitudes, etc.— they grow *new* motive generators. *Motive generator* is a concept proposed by Sloman (1987) and developed by his Cognition and Affect (CogAff) Project in response to the problems of understanding autonomous agency (Sloman 1987, 2008; Beaudoin 1994). Motive generators are mechanisms that tend to produce evaluations, wishes, wants, goals, etc., that may be selected for consideration or physical action. To my knowledge, the concept of motive generator has not previously been explicitly applied to the problems of transfer of learning (Haskell 2000); though I take

---

[1] I include psychological (e.g., affect) theory in cognitive science if it can be expressed as information processing.

[2] The term knowledge resource means a conceptual artifact—such as a paper, book, podcast, audiobook, video, design, illustration, web page—that conveys knowledge. To process knowledge (resources) means to read, view, listen to, converse about, build knowledge with, discuss or otherwise "think" about a knowledge resource.

its applicability to be implied by the CogAff schema (Sloman 2008). The concept suggests a new, mechanistic[3] way to interpret, answer and spawn new problems from the question raised by Bereiter and Scardamalia (1993), "What motivates the process of expertise?"

Bereiter and Scardamalia (1993) coined a distinction between two senses of the word expertise. *Crystallized expertise* is the hard-earned ability to solve problems and perform at a superior level. *Crystallized expertise* is the result of abilities that are applied by people as they strive to become more competent. They refer to these abilities as *fluid* expertise. The terminology parallels the distinction between crystallized expertise and fluid intelligence. White (1959) coined the term "effectance" to refer to motivation for competence. In Beaudoin (2013b), I updated the concept to reflect intervening progress in cognitive science. Effectance now is the (often implicit) motivation to engage in activities that tend to help one solve problems at the edge of one's competence and to push the boundaries of one's understanding of important (and increasingly complex) problems, solutions and objective knowledge. Effectance is an example of a subtle but powerful evolutionary concept developed by Sloman (2009): *architecture-based*motivation. A key idea here is that motivation may be neither conscious nor unconscious, but simply implicit. The concepts of effectance and fluid expertise help one to understand how some become and remain (crystallized) experts while others become, at best, nonexpert specialists.

This chapter in particular addresses the expertise of knowledge workers and life-long learners who increasingly use technology and objective knowledge in their pursuit of personal excellence. While this is not a precise demographic, targeting it allows one to study deep understanding and those who strive for it.

Experts develop highly tuned abilities and tendencies to detect and repair gaps in their understanding, cognitive infelicities, and cognitive opportunities. In the context of the CogAff schema (Sloman 2008), it is natural to suppose that experts develop innumerable, finely tuned motive generators. These mechanisms are constantly responsive to important problems of understanding, which they perceive in a valenced way—meaning that the perception inherently disposes their host to engage in problem solving (Beaudoin 1994). I elaborate this point below.

In general culture, experts are often placed on a pedestal whereas students are frequently seen as deficient. Yet to develop competence one must seek situations that reveal one's ignorance. One must perceive problems of understanding as *enticing opportunities* to better understand. The concept of fluid expertise applies to Marvin Minsky's remark: "No matter what one's problem is, provided that it's hard enough, one always gains from learning better ways to learn" (Minsky 1986). In this vein, the current chapter, though largely conceptual, aims to help experts extend their own excellence as they process knowledge and knowledge resources.

The contributions that this chapter makes to the latter, practical objective, are (1) to characterize the goals and challenges of expert learning, i.e., its requirements

---

[3] "Mechanism" here does not refer to a physical or biological layer, but virtual machine layers (Sloman and Chrisley 2010; Sloman 2010a). It does not preclude but underpins teleology (Boden 1972).

and (2) to succinctly specify how these objectives can be met with upcoming, new technological developments that expand and leverage concepts, principles and findings from cognitive science, namely annotation systems and *productive practice* systems.

12.3 Purposes of Productive Learning from Objective Knowledge

To understand the role of motive generators in expert learning and to enhance such learning, one must consider the manifold aims of learning.

A philosophical preamble is required. Bereiter (2002) argues compellingly that to understand understanding, it helps to divide the world into three (Popper and Eccles 1977; Popper 1979): World 1, the physical world; World 2, the psychological world; and World 3, the world of artifacts, of which only knowledge is central to my chapter. Here, I refer to conceptual artifacts, i.e., designs, theories, models, prescriptions, etc. as "objective knowledge" and to their descriptions as knowledge resources. Understanding lies not in *one* world. Rather, it consists of a relation between a knower (World 2) and objective knowledge (World 3). This relational concept of understanding has many benefits; e.g., it allows one to improve the already potent concept, *knowledge gap* (Vanlehn et al. 1993).

Here follow some overlapping categories of purposes that experts may have when processing knowledge resources. The list glosses over superficial and transient purposes of learning from knowledge resources as well as the practical (psycho-, socio-, and economic) consequences of learning. I focus instead on goals of understanding and of personal enhancement, which are not all easily researched experimentally.

- To understand and work formal (statable) knowledge. To characterize understanding as a relation between knower and an object of knowledge, one needs to take stock of the different types of objective knowledge. Sloman (1978) provides a taxonomy of scientific knowledge (implicit in science's aims). It includes many of the following types of *objective* knowledge: problems, concepts, symbolisms, languages, methods, designs, vocabulary, etc.; real possibilities; correlations, contingencies, explanations of (known) possibilities; limitations (laws, strict principles); explanations of limitations; analyses, criticisms, assessments, etc. Understanding (recursively)[4] requires an understanding of the problem that the objective knowledge is meant to solve. Some objective knowledge has a design. As such, to understand it requires knowing its structure, whether and how the design solves the problem, and how well it solves it (Wertheimer 1959; Perkins 1986). Understanding is neither an all-or-none nor a scalar concept.[5] A thorough understanding

---

[4] Space does not allow me to demonstrate that the circularity in this concept of understanding is virtuous.

[5] Sloman admonishes readers and listeners to beware of the tendency to falsely assume that they are dealing with continuity as opposed to richly structured discontinuous spaces (e.g., Sloman 1984)

involves understanding the space of possible requirements, other designs, and implementations and the relations between these levels.[6] There are other types of knowledge, e.g., regarding the *content* of the world.

- To develop implicit understanding. This involves an ability to work with knowledge, to make predictions and counterfactual statements without necessarily being able to formulate it explicitly or verbalize it to others. (See Karmiloff-Smith (1995) and Sloman (1985) for additional representational subtleties.)
- To develop skills and mastery. This entails abilities to solve problems and achieve goals (cognitive and other).
- To develop episodic, historical and narrative knowledge and abilities to utilize it (e.g., knowing which stories are pertinent to which situations and being able to tell them appropriately).
- To understand norms (standards, processes, etc.) (Ortony et al. 1988), to assess, them, to select them and to regulate one's behavior according to them.
- To develop attitudes towards objects (tastes, likes, dislikes, etc.) For example, one might want to like an unpleasant colleague or dislike salty food.
- To develop habits and propensities. High-caliber knowledge is wasted unless one tends to think and act in accordance with it when it is applicable.
- To develop more abstract cognitive-behavioral dispositions that combine the above. For example, one might want to become more *resourceful* after listening to it; or acquire thinking dispositions, such as to think broadly, rigorously and systematically with the knowledge.

This taxonomy is not complete. (Chap. 2 of Beaudoin (2013b) provides an enhanced and more detailed taxonomy.) It shows that there is a variety of top level goals and types of learning, not to mention the innumerable instrumental ones. Affect is doubly involved in the foregoing list. Learning from objective knowledge is not just a matter of developing "declarative" and "procedural" knowledge.[7] First, affective change is an abstract class of specific learning outcomes towards which one may strive. Second, each of these types of knowledge inherently involves affective change, e.g., developing new motives.

Now here is a major challenge for cognitive science. How can one help experts respond to information such that they can achieve these manifold learning objectives? Transiently comprehending knowledge as one processes a knowledge resource—e.g., about conceptual analysis, emotions, attitudes, resourcefulness—is not a significant problem for the expert. But achieving lasting benefits from it—e.g., to develop the skills of conceptual analysis, to apply potent theories of emotion when solving problems (i.e., achieving broad transfer), to develop desired attitudes towards one's partner, or to actually become a more resourceful person—poses a collection of problems for the individual and cognitive science, particularly in the light of the challenges described in the next section.

---

[6] Thus, the designer-stance (Sloman 1993a, b) to the mind can be applied to other objects of understanding. This is consistent with the elaborate concept of understanding in Bereiter (2002). Problem-centred knowledge becomes requirements-driven knowledge.

[7] Sloman (1996; 2011) undermines the distinction between declarative and procedural information.

New productivity concepts are required in order to improve or supersede the hoary concepts of *active learning* and *deep learning*. Productive learning from objective knowledge involves the kind of productive thinking expounded by Max Wertheimer—it entails understanding. It involves the production of manifold psychological dispositions, abilities, and underlying mechanisms, as opposed to merely the development of content in long-term memory (if there is such a thing) or merely skills and abilities to perform. Cognitive productivity is an optimization of effectiveness and efficiency that involves dispositions to think in a manner that is deep, open-minded, aware, systematic, broad, rigorous, creative, curious, strategic, understanding-driven, and sensitive to unfolding opportunities and context (see Toplak et al. (2012) for a taxonomy.) The *internal* products of productive learning are new mechanisms (such as new motive generators) and configurations of these and existing mechanisms.

These cognitive productivity concepts are aimed at addressing theoretical and practical problems of transfer and satisfaction of personal as well as extrinsic criteria. Theoretically, they are to be developed to address problems of understanding for cognitive scientists (understanding transfer). What must happen during and after processing a knowledge resource in order for the mechanisms to surpass one's former self to grow? Practically, they are tools to focus expert learning, to ensure that more of what is temporarily comprehended is understood, remembered and applied in the long run.

The manifold purposes of learning call for a conceptual understanding of affect and its role in cognition and transfer. Researchers might turn to practical psychological literature on developing affective states (see Sect. 6.3). But psychologists must also revisit information-processing theories and conceptual analysis. More research is required to understand affective processes in expert learning and to develop better practical suggestions for experts. Below, I begin to specify a form of practice, which I call *productive practice*, that is aimed at supporting productive learning from knowledge resources.

Understanding how to facilitate productive learning is of great significance to society. It may allow (groups of) experts to better exploit the opportunities (and face the challenges) posed by the knowledge age.

12.4 Challenges to Productive Learning

In order to support the processes of expertise, one must understand the modern challenges to productive learning from knowledge resources. Expertise does not make one invulnerable to learning challenges; moreover, it does present problems that are distinct from those faced by most students (e.g., cognitive aging) and specialist nonexperts. In this section, I discuss some of these technical and psychological challenges. The technical ones could be alleviated by cognitive productivity software such as an annotation system and a productive practice system. The psychological challenges could be addressed by cognitive productivity workflows and concepts.

The knowledge economy is extremely competitive. Jobs are often scarce. The amount of information that one must process is soaring. Many IT workers, for example "never feel they have enough knowledge for their jobs" (Westar 2009). Yet reading is not enough. Expertise demands productive learning that can be used for progressive problem solving. Determining what to learn is challenging—distractors abound and the future is uncertain. Cognitive aging (Craik and Salthouse 2008), parenting and commuting bring problems and opportunities.

It is no wonder that "productivity software" sells well. However, the category *cognitive productivity* software had not previously been adequately articulated. Cognitive productivity software would be designed specifically to help users achieve the types of objectives listed above while achieving other, extrinsic goals. Some applications could be adapted for this (e.g., outlining and diagramming tools). However, two important categories of cognitive productivity software are not available commercially.[8] One has to do with annotating; the other with practicing. Ultimately, one's computers, tablets, smartphones, etc., need to be an integrated productive learning ecosystem including these two types of applications.

12.5 Information Processing

Here I describe information processing challenges faced by experts.

Technology serves objective knowledge in various formats: PDF files, e-books, web pages, emails, videos, audiobooks, podcasts, etc. Meanwhile, paper has not gone away. Internet-enabled applications connect experts to each other.

Each expert must develop his own cognitive workflows using a hodge-podge of software. Even for one type of information (ebooks) users may rely on several different applications (e.g., Kindle® and iBooks®.) An expert needs to annotate, organize and link information together in a systematic, powerful, and coherent way to rapidly re-access and use it. Each application has its own interface and limited set of capabilities. One cannot yet even uniformly highlight or precisely link arbitrary text (let alone audio and video) from the various formats listed above. Tagging text is not possible in most applications. Neither is annotating in a general fashion.[9] Such links would need to be robust under knowledge resource changes. Furthermore, experts must either create their own glossary management tools or manage glossaries spread across numerous servers and knowledge resources.

Even today, publishers and major software vendors have not adopted schemas (Vlist 2002) or implemented software that would allow users (and software) to query

---

[8] Existing annotation, practice and "cognitive fitness" software applications address narrow requirements (e.g., types of document and types of learning outcomes) Beaudoin (2010a, b, 2013a).

[9] By general annotation, I mean to link any information item accessible from the local host to any existing or self-authored one (e.g., a new note). An example of general annotation is to link a paragraph in an iBooks document to a snip in an email message. My colleagues and I have created several personal learning environments with extensive annotation capabilities, e.g. Beaudoin and Winne (2009).

a schema-compliant knowledge resource for information, such as its table of contents (which ought not to be limited to the few levels that typical books provide but should include all headings), index and bibliography. Knowledge workers must process each knowledge resource to obtain this information. This is unnecessarily tedious. They should be able to issue a command to have the information presented to them—in the style sheet of *their* choice.[10] This information could be extracted and form a starting point for *their* detailed meta-documents, i.e., documents about the resources they are reading,[11] with powerful outlining,[12] editing, tagging, search, referencing, and productive practice integration.

Adding to the expert's woes, previously studied knowledge resources are distributed across multiple platforms: e.g., a computer, ebook reader, tablet and smartphone. The expert must devise a scalable way to sync his knowledge resources, meta-documents, etc. Of the exabytes of content available to the expert, of the subset one "reads," what matters most is that which one has read carefully and one's thoughts about it. Yet due to technical issues, even the expert likely does not properly annotate most of the noteworthy electronic documents he reads.

It appears that even most technically savvy experts have not yet developed optimal solutions to these problems. Moreover, many fail to use powerful cognitive productivity software that does exist (e.g., outliners and diagramming tools). Having focused on cognitive productivity problems since 2001, I believe these issues are tractable. With well-articulated requirements, expertise in cognitive science and an engineering approach, adequate solutions (software and workflows) can be specified and implemented. The shallow, defeatist meme that the Internet has "rewired our brains" and that one is doomed to light learning is easily refuted (Pinker 2010) and will hopefully become extinct as better solutions are developed and disseminated.

A landmark literature review showed what one would expect, namely that as experts read important paper documents they (often zestfully) seek overall meaning, make and adjust predictions about the problems and content, categorize, and assess— while leveraging their prior knowledge (Pressley and Afflerbach 1995). This they do flexibly, often writing as they read. Well before the Internet, knowledge workers developed reading and annotating schemes to deal with overflowing information (e.g., Selye 1964).

Annotation software should allow experts not merely to highlight text but to categorize and extend information in their terms. Examples of fine-grained annotation tags an expert might use to categorize text, images, audio and video segments include: purpose, thesis, major proposition, ancillary proposition, term, concept, definition, criteria, question, author, hypothesis, premise, conjecture, methods, results, data,

---

[10] A simple example of this would for readers to be able to choose for any scholarly document whether to view it in MLA, APA or some other format. This entails the separation of data from its presentation, as is commonly done with XML and cascading style sheets (CSS).

[11] Reading is just a special case. This applies to processing knowledge resources in general. Beaudoin (2013a, b) explicitly deals with knowledge processing in general.

[12] Why should the annotator need to switch to a limited editor in a special-purpose annotation tool? An annotation system could easily leverage the user's preferred word processor, outliner and diagramming tools. This is in the spirit of Poplog Ved, emacs and OpenDoc.

findings section, key argument, warrant, "I disagree", interesting, "I do not understand" (i.e., knowledge gaps), to do (follow up on, reread, etc.), irrelevant, comment, learning questions. A configurable color scheme could enhance this deeper coding.

Users should be able to quickly (within 2 s) locate any knowledge resource they have annotated. Given such a resource, one should be able to rapidly list or locate one's annotations by category (e.g., to find the thesis, extract the technical terms, or their points of disagreements). Navigation between comments, the annotated resource, one's personal glossary and related instillers (as defined below) should be very rapid.

The absence of this functionality, workflows and skills, hinders cognitive productivity in ways that experts may not explicitly realize but would easily understand if the tools were made available to them. Even these tools would not be enough for experts to meet current demands—something more transformational is required.

12.5.1 Productive Learning

In this section, I describe psychological and technical challenges to deriving deep benefits from information. Whereas the empirical study-strategy literature has focused mainly on formal paper/pencil education with normal-range students (as opposed to expert self-education with technology) (Flippo and Caverly 2009), some of the educational psychology findings are relevant to our interests here, e.g., about metacognition, self-regulated learning, and self-testing.

Many students are quite susceptible to the following types of "illusions of competence" (Karpicke et al. 2009) as they process documents. They

- fail to recognize that they have not properly comprehended what they have read (their knowledge gaps) (Karpicke et al. 2009). (The illusion of understanding.)
- overestimate the likelihood that they will remember what they have read (McDaniel et al. 2008). (The illusion of remembering.)
- fail to predict that they will not be able to solve diverse problems with the objective knowledge they have processed because they will not do what it takes to ensure transfer—and to adapt their processing accordingly. (The illusion of transfer.)

Whereas the illusion of understanding is much less a problem for experts, I suspect that experts are not immune to the latter two illusions. Further, while if prompted, experts might make better judgments of learning than college students, they might not be prompting themselves sufficiently (e.g., skimming too much). Given the availability of information, the demands to process large amounts of it, and the state of technology described above, perhaps many experts are not spending enough time ensuring they can utilize the most potent information they "consume" (see also Schopenhauer 1841, Chap. 3; Lamontagne 2002).

Students tend to overestimate the effectiveness, for both understanding and remembering, of re-reading documents or applying ideas (e.g., in open book exams and assignments); conversely, they under-estimate the effectiveness of being tested (closed book) (Karpicke and Blunt 2011). I suspect that experts are also susceptible to this error. If they were properly informed about the implications of these findings and how to leverage them with software, they might adapt their learning strategies and improve their cognitive productivity (i.e., they might more deeply understand, recall and apply what they learn, while consuming less time overall).

The market has yet to produce a second important class of cognitive productivity software. Experts likely have answers to questions such as: "What applications do you use for: writing email? Developing a spreadsheet? Composing a document? Browsing the web? Reading PDF files? Drawing a diagram?" But if you ask: "What software do you use when you want to not only read but learn something (i.e., turn information into your own knowledge and mindware)?" an answer is likely less forthcoming. If you were to further explain that the software should help to achieve the manifold purposes of learning described above, there may still be no answer. These requirements call for a productive practice system with document annotation capabilities. There are many principles of cognitive science that could inform the development of cognitive productivity software and workflows which would have significant impact on knowledge workers—particularly if, like many other applications, they shipped with operating systems.

12.6 The Role of Motive Generators in Productive Learning

> Cognitive psychologists are often accused of ignoring motivation. A more generous appraisal would be that they honour the principle if you do not have something worthwhile to contribute on a topic you should refrain from speaking. Most of what psychologists of any sort have to say about motivation is warmed-over common sense. The part that is not common sense involves the brain, but it is at such a basic level that we cannot expect it to be helpful in distinguishing experts from experienced nonexperts. (Bereiter and Scardamalia 1993, p.101).

There is truth to Bereiter and Scardamalia's view. However, I take issue with the idea that motivation theory must either be folk psychology or biology. Moreover, folk psychology ought not to be ignored—let us not forget that Einstein's theory of relativity involved his analysis of ordinary concepts. Psychology, when its research methods are divorced from conceptual analysis, does not do justice to the affective lexicon of the English language (Ortony et al. 1987). Alas, conceptual analysis sometimes goes astray. For example, a problem with the influential book, *The Intentional Stance* (Dennett 1987), is its coarse reduction of affective states to the concepts of belief and desire. That book is not unique in over-emphasizing prediction and neglecting the importance of understanding; for understanding is a relation that almost always requires dealing with the object's inner structure (in the tradition of Kant, Wertheimer, Bartlett, and design-based AI). One of the problems with typical approaches to motivation is that they are too focused on explaining *why* individuals do things, rather than *how* the mind enables them to generate, process and pursue their goals. It tends to reduce motivation to scalars (mainly intensity), thereby ignoring rich structures and information-processing underlying motivation.

There is an over-emphasis on data collection and an apparent failure to recognize the key insight of (design-based) cognitive science: One can only discover the *actual* mechanisms of mind by specifying what the mechanisms, architecture and control sub-states (Sloman 1993b) might *potentially* be in relation to the requirements that they must satisfy. This is not to say that the designer stance guarantees rapid success, but it raises germane questions.

The CogAff project's approach couples the designer stance (McCarthy 2008; Sloman 1993a) with conceptual analysis (Sloman 1978). The designer stance encourages one to develop conjectures about information processing mechanisms that are layered on the brain and other physical devices.

The CogAff theory assumes that the mind is perpetually generating (and not merely deriving through means-ends analysis) motivational states (motives). We assume a human mind contains a multitude of motive generators. These mechanisms perpetually monitor internal processes, events and states and respond by creating motivational states (motives), i.e., control states that incline one towards affecting the (internal or external) world in some way. Humans may have fewer than ten physical perceptual modalities, but they have countless internal monitors. Motives underlie our wishes, wants, desires, whims, preferences, etc. The fact that the mind is inhabited by these mechanisms has never ceased to fascinate me, ever since I first read about them in Sloman (1978). In my opinion, understanding motive generators and how they develop is as fundamental to cognitive science, writ large, as understanding force and energy is to physics.

This section addresses one aspect of the following question: In what way does a mind need to change in order to be able to apply what one has previously learned? In other words, what must happen in one's mind such that one applies this information far from its original context? In this section, I propose that this involves creating and configuring motive generators. These mechanisms produce motivational states to respond to *future* applicable situations. An important goal of (self) education must implicitly be to fine-tune these motive generators such that they create the right motivational states at the right moment. Another goal is for management and meta-management processes to properly process these motives when they surface.

I will provide examples of three types of transfer targets: skills, understanding and attitudes. My answer makes use of concepts developed by Sloman (1987) and other members of the CogAff project that were based on Simon (1967) as were Frijda (1986), Oatley and Johnson-Laird (1987). The CogAff project sought to understand, amongst other things, motive processing requirements, architectures, mechanisms and representations. Productive learning impacts on each of the latter. In particular, it must affect motive generators and management processes (Beaudoin 1994).

My conjecture is that as one learns productively, one's mind creates and fine tunes a potentially large collection of motive generators. These mechanisms become tuned to respond to specific opportunities and problems that pertain to the information in question, at various levels of abstractions. Many of these monitors are sensitive to problems in the solutions that other mental mechanisms generate, or that other people have generated, or that are explicit or implicit in conceptual artifacts.

A key design problem for transfer is to decide how the insistence of these motives is to be determined. In previous work we characterized insistence as heuristically related to the importance and urgency of motives (Beaudoin 1994; Sloman 1987). I have previously analyzed the components of goals (Beaudoin 1994). Another factor that contributes to determining the insistence of motives that are derived from mastering objective knowledge is the perceived usefulness of the particular objective knowledge for a range of problems (Beaudoin 2013b).[13] This is akin to the usefulness of any tool (conceptual artifacts are tools). An important part of the development of expertise (i.e., of fluid expertise) is being able to recognize the efficaciousness of knowledge for problems that one might encounter. This should influence the effort that one expends on mastering the knowledge. Repeated, elaborate use of knowledge in varied real or imagined situations depends on and influences judgments of its efficaciousness. Using knowledge creates motive generators that monitor for situations in which it might be of future use. An expert mind must do what it can to ensure that the motive generators related to potent tools develop adequately. This will often initially lead to situations in which the resultant (potent-tool related) motives have too high insistence. Compare when one considers using a new kitchen instrument when a different one is more appropriate.

The heuristic nature of asynchronous motive generators (Sloman 1987) has a dark side. For example, knowledge workers are exposed to red herrings—alluring, impotent information that may trigger motives to process it. Motives do not necessarily distract attention; nor do they necessarily lead to external action. For instance, higher level processes can reject a motive until the next time it surfaces. But cognitive processing is an action of sorts, which must be monitored to detect distractions (an example of what we have called meta-management—Beaudoin (1994)).

The importance to productive thinking of detecting what are now commonly called "knowledge gaps" cannot reasonably be disputed. Like many scientific propositions, that experts have a keen ability to detect and repair knowledge gaps is not merely a contingent, empirical truth; it is analytical. Experts continue to develop new motive generators to detect new types of knowledge gaps. They create motives to repair the gaps. Frequently, if one can detect a problem of understanding in a valenced way, one is most of the way to solving it; one merely has to "dare to think." A challenge is to detect the knowledge gap on the basis of the applicability of the objective knowledge with which one is already sufficiently familiar.

Max Wertheimer discussed the perception of knowledge gaps in his own terms. He emphasized that productive problem solvers become keenly aware of discrepancies between the "requirements of the problem they face" and their current "view" when it lacks adequate "penetration" or "clarity." Of course, experts do not always recognize their flaws. Still, it is important to keep knowledge gaps in mind when thinking about learning. In this chapter I focus on situations in which the expert already has acquired much of the information that he needs to address a problem. This enables

---

[13] This is related to *promisingness* in Bereiter and Scardamalia (1993) and *potential* in Sinclair (2006).

one, subsequently, (somehow) to detect gaps in one and others' knowledge, which can lead to more progress.

One can ask numerous questions from the foregoing, such as what motivational mechanisms develop, enabling one to detect increasingly sophisticated knowledge gaps. What are the possible internal differences between the case when one has understanding of some objective knowledge and (a) yet one tends to fail to apply (transfer) it and (b) when one does apply it? One possibility is that a motive generator actually fires in response to a knowledge gap in (a), but the motive is not sufficiently insistent to attract attention. What are the different ways in which this can happen? In answering these questions, we should not limit our attention to a simple layered architecture, in which a layer of motive generators is connected to one interrupt filter layer, connected to one management layer and a meta-management layer. Motive generators (i.e., monitors) are spread throughout a human-like mental architecture. There are monitors of monitors (meta-monitors). Failure to transfer could be due to failures in meta-monitoring.

The English lexicon contains a much larger number of words denoting negative affective states than positive ones (Fredrickson 1998). This may be pertinent to the problems of transfer. In the much prized state of "flow," the expert proceeds smoothly, solving one problem after another, gracefully dealing with setbacks. Progress is a rougher ride; expertise involves pushing the boundaries of one's competence, which means that one must face what are potentially annoyances or worse. Experts require cognitive zest, to put a positive spin on the difficulties they encounter. A setback, a difficulty, a bug in a computer program, a relationship that is not quite right—these are often opportunities to improve one's knowledge. But for many experts—perhaps for most—negative affect is present in otherwise optimistic states before the hurdles are overcome on the road to further excellence. Perhaps this fact about the English affective lexicon reflects the human condition and entails that learning produces more motive generators to detect impasses than to detect (and enjoy) progress. (How could this be determined?)

In the following subsections, I will provide examples of the role of motive generators for three types of knowledge: skills, dealing with a theoretical problem of understanding, and attitudes. The examples are somewhat Escherian in that they directly involve matter developed by Sloman and leveraged in the writing of this chapter. These are not explanations but loglines, scenarios and allusions for further elaboration, requirements analysis and design exploration.

12.6.1 A Cognitive Skill Set: Conceptual Analysis

In this section, I will provide an example of the role of motive generators in someone who is skilled at conceptual analysis. I could have chosen any other cognitive skill. However, this one is particularly apposite to this collection—given the emphasis Sloman has placed on conceptual analysis—and to this chapter, given that conceptual analysis demands and promotes deep understanding. Moreover, as one masters conceptual analysis, motive generators are created that ought to be active in a wide variety of verbal problem solving situations.

Briefly, conceptual analysis allows one to improve one's understanding of concepts that are intuitively understood by many intelligent speakers of a natural language. English, having the largest lexicon of all natural languages, is a particularly fertile language for this activity. Sloman (1978, 2010b) has provided detailed descriptions and suggestions about how to perform conceptual analysis. Here I describe the kind of mind that has studied these readings in depth, thought about them and applied them.

Many university students—who may be strong in other areas—find it difficult despite instruction therein to understand and perform conceptual analysis. (What distinguishes those who get it from those who don't?) Yet conceptual analysis involves cognitive skills that are essential to day-to-day knowledge building, academic pursuits, and scientific knowledge building (Wilson 1963). Although conceptual analysis normally takes informal knowledge as a starting point, its inputs can include objective knowledge and it can develop new concepts. It bears repeating that Einstein's theory of relativity stemmed from a conceptual analysis of space and time. My own Ph.D. thesis contains a conceptual analysis of goals and motives. Scientific progress, particularly in the social sciences and psychobiology, is often held back by lacunas in conceptual analysis.

Reading conceptual analyses, reading about conceptual analysis, performing conceptual analyses and receiving feedback on one's analyses modifies one's mind in important ways. In particular, one grows new motive generators. One comes to monitor what one reads, thinks, and says, detecting conceptual infelicities, opportunities, and noteworthy facts, such as the following. ("One" might be oneself or another. The citations below are mainly to documents that comment on the respective issue.)

- This concept is (potentially) very (im)potent (for some set of problems).
- This term, concept or distinction is new (to some or to all).
- This concept fills an important knowledge gap (of mine, of the community).
- This is a polymorphic (Sloman 1978, 2010b), cluster or suitcase (Minsky 2006) concept. What meaning is one using? What are the different meanings? What words should one use for the differences?
- One is switching (explicitly, equivocating or havering) between meanings of a term.
- One has failed to make important distinctions. Reality is not being "cut at its joints" (Stanovich 2009). The same term is being used for different concepts. See Lakatos (1980)'s elegant progression of distinctions.
- One doesn't understand something important about this concept (knowledge gap).
- A category mistake has been made (Ryle 1949).
- This definition is circular (viciously, acceptably, inevitably (Sloman 2011)).
- This definition is misleading (e.g., because it rules out (or in) cases that it ought to include (or exclude) or it has some other infelicitous implications).
- This concept has been vitiated by this definition.
- This is a false dichotomy.

- A structural concept is being inappropriately reduced to a scalar one (Beaudoin 1994).
- This concept has self-defeating semantics (Sloman 2010b).
- This concept is emblematic of a degenerating research programme (Lakatos 1980).
- This concept lacks explanatory or generative power.
- The logical geography of this theory is a particularly small part of the relevant logical topography (Sloman 2010b).
- The term is being used differently from its accepted, expected or referenced meaning.
- This high-level concept is being treated as a basic-level one—in a referent-centred rather than problem-centred way (Bereiter 2002).
- The resource's lexicon (e.g., based on the text, index, or glossary) is not sufficiently rich to address its objectives (an example is the index in Gladwell (2000)).
- The limits, boundaries or conditions of applicability of this concept are unclear or troubling.
- The author of a computer program errs in failing to apply a powerful structuring concept. For exmaple, Leach (2011) showed that Beck's 2002 use of incremental design overlooked the concept, *bag*.
- This distinction adds no value (principle of parsimony has been violated.)
- It is (not) worth arguing whether the definition is right in this case.
- A new concept, taxonomy or language is required to address these problems.

As these examples and White (1964) suggest, the types and time courses of realizing are varied. An expert does not normally apply a checklist of criteria against a document he or she is reading. A large collection of special purpose monitors are at work in the expert mind, in parallel, observing records of high-level mental processes (related to the resource and objective knowledge). They generate motives of varying insistence, which may (or may not) influence the reader's information processing. The motives are not necessarily goals or intentions; they are often merely valenced descriptions that may lead to the formation of specific goals. These motives involve a cognitive itch,[14] e.g., that something is wrong. They will normally require attention. What representations are useful for these motives? Karmiloff-Smith (1995) points out that there are many different representational types, not just "explicit," "implicit" (see also Sloman 1985; Beaudoin 2013b). How do these mechanisms interact with management processes?

Conjecture: The expert-like novice in a domain must somehow grow these motive generators with respect to whatever skill being developed from objective knowledge. In order to develop, these motive generators must frequently drive cognition. That is,

---

[14] The concept of cognitive itch needs to be articulated in designer terms, to surpass the limitations of conceptual analysis. The term "cognitive itch" has been used independently by Beaman and Williams (2010). The "itch" I am describing is a state in which one detects a cognitive infelicity and wants to do something about it (whether or not the motives surface or one deals with it). Beaman's "itch" is better renamed and classified as a cognitive perturbance (Beaudoin 1994), which is one of many possible states that a class of motive processing systems can generate, rather than merely as an arbitrary phenomenological state.

their motives need to be sufficiently insistent to surface and periodically spur problem solving. Otherwise, they will lie fallow and the result will be inert knowledge—skills will not develop. Once motive generators have been sufficiently active, they will acquire functional autonomy (Allport 1961; Beaudoin 1994) and, barring aberrations or supersession (how?), will remain active indefinitely. One will continue to notice and be irritated by conceptual infelicities. This analysis supposes that mastery of cognitive skills creates many affective states.

12.6.2 Detecting Conceptual Infelicities in Talk About Grief

This section uses a scenario to illustrate the role of motive generators in applying previously developed understanding of two complementary theories of emotion—Sloman's perturbance theory (Sloman 1987) and Ortony et al. (1988) cognitive structure theory. The likelihood that (and ease with which) knowledge can be applied depends on a number of factors including the criteria for understanding described by Bereiter (2002) and Beaudoin (2013b). The scenario deals with a case in which one had acquired a certain mastery of the two theories but allowed it to lie fallow (inert) until critical developments led to its application—some motive generators became productive again.

Suppose the wife of a happily married software engineer, somewhat familiar with the aforementioned theories, dies in a car accident. His grief subsides; but he is sufficiently troubled (and intrigued) by the remaining perturbance to wish to understand it and to better control his mental processing.

In discussing his situation with loved ones and a psychologist, he becomes annoyed by the welter of affective concepts that prevent him from thinking clearly about his experience. Even the psychologist's concepts and terms do not seem right to him.[15] Hence, our grieving engineer becomes motivated not merely to understand his grief, but emotions in general. He (correctly) feels that perhaps this understanding will help him take a healthy distance from that which causes his negative affect and improve his mood.

He is thus motivated by the *efficaciousness* of the theories he had encountered which now acquire *additional* value from being instrumental to his goal of equanimity. (How did this recognition trickle down to his (cognitive) motive generators?) His conversations trigger a cognitive itch in him. Questions arise somehow from his previous understanding of relevant theories.

- Psychologist: "Your anger is real and it must go somewhere" (Worden 1991, p. 43). Engineer: "But emotions are not substances that can be shunted. If anger is "real," what *is* it really?"
- Psychologist: "You are not consciously aware of your feelings" (p. 44). Engineer: "In my understanding, feelings may be fleeting, low level, unverbalized or

---

[15] I have suggested that clinical psychologists in particular should be trained in conceptual analysis (Beaudoin 2013a, b) and the designer stance (Beaudoin 2013b).

unacknowledged but not unconscious. It is not my feelings that I need to better understand, but the mechanisms in my mind that produce my feelings, thinking, deciding, planning, assessments, my manifold appraisals, etc." See also (Worden 1991).

- Psychologist: "Perhaps you feel guilty because you are not experiencing enough sadness?" (p. 45). Engineer: "You are over-emphasizing feelings and neglecting the cognitive structure of emotion. How can I understand feelings without reference to a taxonomy of emotion?"

While the engineer's responses are sophisticated and interesting in themselves, our issue lies in the interval between his interlocutor's statement and the engineer's articulation of his cognitive *concern* or *itch*. In a brief but complex moment, he *becomes* genuinely dissatisfied with his interlocutor's statements. He notices that something is wrong. Through his prior learning, he had understood important ideas about affect. But this knowledge had remained relatively inert. Faced with a pressing need to understand his experience he became sensitive to his interlocutors' and his own ignorance. His prior learning had established motive generators that allowed him, years later, to detect possible knowledge gaps in himself and others, whereas his interlocutor seems oblivious to the problems.

To understand this example of transfer, one must address designer-based questions (beyond conceptual analysis) such as:

- How does understanding later lead to cognitive itches?
- What information processes constitute the "cognitive itch" and elicit questions? That is, the aforementioned moment needs to be mechanistically described from the designer stance.
- How were the monitors established originally?
- What are the monitors connected to? What are their inputs? What precisely are the monitors looking for?
- What are their outputs?
- What might have happened internally, in mechanistic and architectural terms,[16] such that the engineer would have applied and developed these motive generators ever since he "learned" the theories? (i.e., had the knowledge not remained inert).
- Before his loss, did incoherent talk about emotion trigger similar but less insistent motives? Or were they not generated at all?
- What is the mechanistic (not merely the "content") difference between the engineer, as he notices an infelicity, and his interlocutor who is oblivious to it?
- How did his monitors divert attention?
- What are the dimensions of variation of, and the structural variations between, the various motive generators involved in detecting conceptual infelicities?
- How can such motive generators be established such that they persist almost indefinitely?

---

[16] Notice that some of the epithets that a designer uses in his quest for understanding are a modernization of Wertheimer's "internal structure" talk. The designer is concerned with internal functional architecture.

- How can new understanding dismantle or attenuate motive generators that are no longer relevant?

These are all deep questions whose answers will shed light on understanding and transfer.

12.6.3 Developing Attitudes

The scientific case for motive generators in attitudes is not difficult to make, if one accepts that attitudes are "dispositions, or perhaps better, predispositions to like some things, e.g., sweet substances, or classical music or one's children, and to dislike others (e.g., bitter substances, or pop art or one's enemies)" (Ortony et al. 1988, p. 328) and that goals involve a motivational attitude towards information (Beaudoin 1994). A state that does not tend to generate motives is simply not an affective state. That opens many questions about *how* motive generators develop and operate as part of the information processing substrate of attitudes.

The main question relevant to attitudes that arises in this chapter is: how can one develop attitudes, and hence the motive generators underlying them, through interaction with objective knowledge? An expert (e.g., outside cognitive science) might read about the role of an attitude and infer that he needs to change his attitude(s). But how is this accomplished? Attitudes, moods, intricate cognitive-behavioral dispositions and beliefs are not all states that one can simply decide to change.

Some psychologists have practical recommendations regarding personal attitude change that may be relevant here. For example, many of John Gottman's recommendations for improving relationships are directly aimed at changing attitudes. Gottman et al. (1999) advise their readers to nurture their mutual fondness and admiration, which requires attitudinal change. For this they suggest taking turns in complaining; not giving unsolicited advice; showing genuine interest; communicating one's understanding; taking one's partner's side; expressing a "we against others" attitude; expressing affection; and validating each other's emotions Gottman.

The desired change is unlikely to occur without extensive, self-regulated practice of the theory (e.g., analyzing bids for connection and developing love maps.) Experts may benefit from guidance on how to modify *arbitrary* attitudes based on any specific, practical, and useful objective knowledge they may encounter. This involves a separation between descriptions of attitudes and descriptions of the means for developing those attitudes. Productive practice, described below, is a strategy of regular elaborative practice and exercises derived from content one wants to master. If practical authors (such as Gottman) are correct about the implicit possibility of changing one's motive generators, and certainly much of clinical psychology makes this tacit assumption, then such a shell may be of use to those seeking attitude change.

The next section tries to shed some additional light on this, and on how skills and conceptual understanding are similarly developed.

12.6.4 Developing Deep Understanding

> I think there is only one way to science—or to philosophy for that matter; to meet a problem, to see its beauty and fall in love with it; to get married to it, and to live with it happily, till death do ye part—unless you should meet another and even more fascinating problem, or unless indeed you should obtain a solution. (Popper 1983, p. 8).

Developing an understanding of objective knowledge and developing concomitant motive generators does not happen instantaneously. It poses a challenge to breadth seekers. The requirements for understanding objective knowledge described by Bereiter (2002) and Perkins (1995) follow a pattern that is similar to the designer stance of Artificial Intelligence (Sloman 1993a); that is: (1) knowing the environment of the object; (2) knowing its requirements or purpose; (3) knowing its structure (design); (4) knowing its implementations; (5) analyzing how the design meets the requirements, and the implementations meet the design specification; and (6) understanding how changes in requirements, designs and implementations relate to each other.

Bereiter (2002) argues that deep understanding of objective knowledge involves using the information to solve deep problems. He emphasizes that building new conceptual artifacts with objective knowledge is important for this. (The designer stance, including building computer programs to develop and test one's understanding, is conducive to this.) From the problem-solving literature, we know that experts use examples in particular ways (VanLehn 1996). We know that expertise often takes a long time to develop. From Boden (1991) and some artists, we gather that creative discoveries usually involve dedicated, and intimate involvement with problems and solutions. Based on his extensive interviews of Albert Einstein, Wertheimer (1959) reports that Einstein was concerned with his great problem for 7 years before making the conceptual discovery about time that led him to write his paper on relativity in a mere 5 weeks (while holding an unrelated day job). Perkins (1995) proposed a useful geographical metaphor for intelligence and expertise: knowing one's way around thinking and domains, respectively, which requires deep involvement with the domains. Experts tend to interact extensively with each other directly or through their resources. The examples used by each one of these cognitive scientists calls for an explanation of the motivation involved therein.

How one asks the question "What motivates the process of expertise?" will determine the answer. If one reads this as "what do people aim to get out of this process?", one might answer "flow" (Csikszentmihalyi 2008)—i.e., that "it feels good"—or refer to some other end or reward to which expertise is an instrument. Also perceived self-efficacy (PSE) has been shown to play a great role in individuals' progress (Bandura 1997); but PSE is an *enabling* factor.

The concept of *cognitive zest* is important for understanding expertise. Cognitive zest includes perceived cognitive self-efficacy towards the classes of cognitive problems one selects. Zest entails PSE but PSE does not entail zest. Cognitive zest includes an additional enthusiasm for solving problems of understanding on the way to solving other problems (such as creating new objective knowledge) and enthusiasm for the tasks that inherently lead to knowledge building. Experts are not constantly in "flow" nor are they necessarily seeking flow. They practice, debug, read dry papers and deal with adversity and setbacks. Winston Churchill, whose enthusiasm is described as such by Jenkins (2002), said of courage what one may say of zest, "[it] is going from failure to failure without losing enthusiasm."

Interpreting the foregoing motivation question in terms of what people seek is a source of degenerating research programs concerning affect. The answers point to surface requirements. People seek things for many motives; and many motives have functional autonomy. Many have argued that we do not tend to things for the pleasant feelings the activities *sometimes* generate (Ryle 1949; Reiss 2000) Something else is at play in *effectance* and the pursuit of excellence.

From the designer stance, one interprets the question very differently. "Where the [motive-generating] mechanism comes from and what its benefits are irrelevant to its being a motivational mechanism: all that matters is that it should generate motives, and thereby be capable of influencing selection and generation of behaviors." (Sloman 2009.) Here, one would ask questions like: How do motive processing mechanisms work to evince and sustain the process of expertise? How do they develop *internally*? How do they satisfy their requirements? What might the architecture, mechanisms and representations of a mind be that sustain progressive problem solving?

Cognitive zest is neither content, data nor a mechanism of the mind. It is a requirement of explanations of experts' information processing that they should do justice to cognitive zest. Even if one cannot create an expert robot mind—expert in challenging environments, that is—without that mind being zestful, zest might still only be a second-order intentional property. But cognitive zest, PSE, and flow, ascribed without knowledge of the workings of the mind, entail something about how the robot (or human) develops motive generators. The mechanisms—not those second-order categories—will explain behavior.

12.7 Learning Strategies for Experts

Given that productive learning requires deep involvement with problems and objective knowledge, how are experts to remain abreast of broad literatures and derive deep benefits from them? There are many problem-centered ways to address this question (and indeed many questions to raise from them). Selye (1964) told his imagined son, "Either read or skim through literature, but do not try to do both." In order to process knowledge productively, one must focus on the most efficacious information, meaning the most useful, high caliber and potent information. Caliber is the objective quality of the information. Usefulness is relative to one's goals, projects and constraints. Potency is the extent to which one's mind can be altered by the information. To become more productive, one needs to gauge how one spends one's time in relation to these criteria and the depth of one's processing. Current time tracking software is unfortunately not optimized for this.

The education literature and practical guides describe many learning strategies (Flippo and Caverly 2009). Even if they were suited to experts learning with technology, experts would have an adaptive decision to make when processing knowledge resources: which of the multiple strategies should they use given a particular learning task? I agree with Pressley and Afflerbach (1995) that promoting fixed-sequence strategies (e.g., SQR-3) is not consistent with how experts read or one ought to read. Still, one can propose specific tools and partial workflows derived from the most potent findings in cognitive science for the experts to choose from.

12.7.1 Productive Practice

I propose the concept of *productive practice,* a deliberate practice analog of progressive problem solving (Bereiter and Scardamalia 1993), a form of deliberate, question-and-answer based learning. It leverages direct and indirect test effects, and several potent psychological principles. It is aimed specifically at experts and expert-like novices with high cognitive productivity demands. It explicitly repudiates rote learning. It aims to promote the manifold purposes of learning sketched above, i.e., deep understanding, transfer, the psychological workings of knowledge building, etc. It is also amenable to automation.

Productive practice involves creating, answering and revising questions about what one aims to learn, remember, understand and master, before, while and/or after one initially processes it; and practicing answering these questions (through time and in an elaborate manner) with software that optimizes the practice schedules (i.e., to minimize effort and maximize the learning benefits). Space does not allow a full description of productive practice (see Beaudoin 2013b). At a minimum it should be noted that the kinds of questions one asks and the kinds of answers one articulates influence the productiveness of practice—e.g., problem-centred versus referent centred (Bereiter 2002). For designing productive practice software, I've introduced the concept of instillers, which are data structures about specific information to learn. Instillers have a type (e.g., generic, vocabulary, procedure, person, event, self-regulation, verbatim document), one or more pairs of equivalent questions and answers, references, links, and other information. Practice is progressive through being integrated in continual learning (and often knowledge building) that enables one to progressively extend one's ability to solve problems while maintaining prior knowledge that ought to survive the test of cognitive time (i.e., would otherwise lie fallow).

The strategy as described here is original in its combined emphasis of (a) the manifold purposes of learning sketched above, (b) regular practice, (c) processes of expertise outside of formal education, (d) technology-laden workflows, and (e) cognitive productivity.

The effects of testing on remembering has been the subject of extensive empirical research. One naturally tends to forget information which one does not practice recalling. Practicing recalling information can suspend forgetting and improve speed of recall. This is not to deny one-trial learning nor the relativity of remembering and forgetting (Roediger 2008). The potency of test effects is overlooked by too many in formal education and expertise, though several researchers are spreading the word, e.g., Karpicke and Blunt (2011), Roediger and Finn (2010).

Roediger and Karpicke (2006b) propose that there are direct and indirect effects of testing. When self-testing is used for rote learning (e.g., with traditional paper or software flashcards) it mainly leverages the direct effects of testing. Indirect effects include, for example, motivating the learner to study and providing feedback.

Self-testing is used extensively, systematically, and successfully by many students; but, despite its potential benefits (Roediger and Finn 2010), after graduation from university, most of these same people (even if they become knowledge workers) do not as systematically engage in such practice. The next section briefly explores why knowledge workers do not and why some of them should.

12.7.2 The (Neglected) Benefits of Productive Practice

The concept of productive practice has not been sufficiently articulated and disseminated. Rather than describe its structure in detail here, I explore why productive practice is not an explicit part of the cognitive toolkit of most experts. Then I list anticipated benefits which can be interpreted as requirements or criteria for assessing this cognitive tool.

- Question and answer practice tends to be confounded with rote learning and repetitive practice.
- Productive practice, theoretically, is not understood as distinct from one of its components, distributed recall practice which itself is often characterized in (nebulous) terms of "memory traces." Yet, there are more types of information, more ways of acquiring information, more ways of storing information, more time scales over which information is processed, stored, and used than trace theory allows. Similarly, the procedural-declarative distinction is a naive dichotomy; for familiarity with computer programming structures suggests manifold types of mixed data, e.g., Gibson (1994). Reproductive memory, for example, can be implemented by mechanisms that can reconstruct explicit forms of knowledge at a later point through stored procedures. Productive practice is not dependent on trace-based explanations.
- Although there are multiple purposes of learning, distributed recall practice is not typically aimed at many of them (but productive practice is).
- Although the test effect has been studied, technology has not been developed to collect pertinent information from hundreds of thousands of users to optimize productive practice software.
- Experts have not been exposed to productive practice workflows that leverage other potent principles and concepts, such as goal setting (Latham and Locke 1991), optimal cue generation (Norman and Bobrow 1979), progressive problem

solving (Bereiter and Scardamalia 1993), elaboration by argument (Wiley and Voss 1999), re-representation, and self-explanation (Vanlehn et al. 1991), PSE (Bandura 1997).

- Some of the current vocabulary for practice is awkward and misleading (e.g., "distributed recall practice," "self-testing," "reviewing," and "flashcards"). New concepts are also required (e.g., instillers, challenges, effectance, meta-effectiveness, linkback, mindware, reconstructible discriminative cues, potency). Beaudoin (2013b) describes these concepts.
- Deliberate practice is implicitly not often considered as pertinent to professional knowledge work as it is to performance sports, performance arts and formal education.
- There may be the perception that practice is for novices, not experts. Yet many experts do implicitly practice (e.g., through teaching, writing, and using their knowledge). Those who frequently present on varied and difficult problems practice de facto and often deliberately.
- The problem of transfers is treated as insurmountable. Productive practice is proactively designed to address the problems of transfer.
- Productive practice strategies, integrated with knowledge acquisition workflows and technology, have only recently been published (Beaudoin 2013b).
- There is a tendency towards over-reliance on external memory aids, including the Internet.
- Educational psychology focuses mainly on formal education and developmental students rather than on lifelong learning and expertise (Mulcahy-Ernt and Caverly 2009).
- Experts might feel they are too busy to practice, not realizing (a) that they can decrease (re) reading time and obtain more lasting benefits with practice (i.e., productive practice is a cognitive productivity practice); (b) mobile and other productive practice opportunities exist; (c) elaborative practice is an important component of expertise that can be systematized.

Productive practice is meant to address all of these concerns. Here are some additional *anticipated* benefits of productive practice.

- The well-documented benefits of distributed recall practice apply to it.
- Unless one's life involves frequent presentations and meetings about target knowledge, without productive practice one might not sufficiently articulate, develop and truly understand potent knowledge.
- Productive practice can be integrated in opportunistic and systematic reading and learning workflows, e.g., mining information gems from knowledge resources and selecting the essential subsets to not only comprehend but master.
- Productive practice allows one to systematically control mastery of objective knowledge in conjunction with one's goal setting practices.
- Productive practice supports multiple learning purposes mentioned above, not merely factual knowledge. For example, if John Gottmann is correct, then regularly answering questions about *bids for connection* within a couple may help members become more mindful and emotionally satisfied (Gottman et al. 2001, pp. 65–69).

- Productive practice addresses problems many people have in how they practice (Guadagnoli 2009).
- Productive practice facilitates the detection and management of (sometimes subtle) knowledge gaps and cognitive opportunities (e.g., to connect or elaborate information) while improving one's current and future judgments of learning.
- Productive practice helps one to develop long-term working memory (Ericsson et al. 1993; Ericsson and Kintsch 1995) of the information that matters.
- Productive practice sharpens one's abilities to ask and answer productive questions, to analyze resources (if it is integrated with a resource processing workflow), and regulate one's learning.
- Productive practice prepares one for future learning (Schwartz et al. 2005).
- Productive practice helps one apply knowledge: one is not merely passively primed with target information but one actively recalls, reconstructs and processes it. One can practice it in different environments, which fosters generalization and transfer.

In accordance with the section on deep understanding above, productive practice requires effort. For most people, it will represent a change to how they process resources with technology. Existing technology has not still been adapted to optimally support productive practice and cognitive productivity.

12.8 Future Research

In keeping with the theme of this symposium, hard problems in the study of cognition, I have raised more questions than I have answered. I conclude by raising the explanatory bar even higher.

Mechanisms underlying the effects of testing are poorly understood and in need of deep explanation. I believe the common expression "test effect" is as much of a euphemism in cognitive psychology as "the gravity effect" would be in physics. Existing conjectures—which the literature refers to as hypotheses—in terms of desirable difficulty (retrieval effort) (Schmidt and Bjork 1992), elaborative retrieval (Carpenter and DeLosh 2006), transfer-appropriate processing (Carpenter and DeLosh 2006), mediator shift (Pyc 2010), and adaptiveness (or rationality) of memory (Anderson and Milson 1989) are a source of cognitive itch in my mind. It is not that they are empirically wrong. It is that they do not describe mechanisms nor are they derived from mechanisms or architectures. As such, although they are pertinent, they do not explain the important phenomena they are supposed to address. Moreover, despite experiments that pit one against the other, they are not all incompatible. In my opinion, their usefulness is at the level of requirements. Requirements are an important part of theory: they are meant to drive designs. Rather than proposing more hypotheses or running experiments to test them, we need to develop designs that explain the basic phenomena and from which hypotheses may be rigorously *derived*. In exploring potential underlying mechanisms and architectures, I expect the distinction between direct and indirect testing effects will give way to a collection of distinctions. The 'test effect' seems to get at something so fundamental that I further suggest, albeit vaguely, that its explanations will be tied to major mechanisms of autonomous agency, i.e., motive processing (including motive generators).

Further, there are two strands of research that are in need of integration. The problems addressed by broad theories of self-regulated learning (Winne 2001) overlap substantially with the problems addressed by the CogAff project (and some other architecture-oriented theories in cognitive science—e.g., Winne (2001). Winne's (2001) statement that "Metacognitive monitoring is the key to self-regulating one's learning" is consistent with the thesis of this chapter that motive generators (i.e., monitors) are deeply involved in transfer. While this chapter focused mainly on motive generators, there are also many questions to be raised about how other motive processes, representations and entire architectures make transfer possible.

In the spirit of Bartlett and Wertheimer, empirical studies on productive practice should not focus on rote learning (e.g., paired associate tasks) but focus on authentic, meaningful, and conceptual learning. When recall is the only concern, "distributed recall practice" is a more apposite term to use (not that retrieval practice is necessarily rote).

Productive practice needs to be specified in more detail than space allows here (but see Beaudoin 2013b). Only then will the anticipated benefits described in the previous section need to be assessed empirically—though many of them are to be expected given that they are based on some of the most well researched theories, principles and findings in cognitive science. Cognitive productivity software is poised to become an important area of application (and development) of cognitive science.

Acknowledgments Thank you to Jeremy Wyatt for the initiative and all the effort in organizing this important event in the history of cognitive science. Thank you to Carl Bereiter, Alissa Ehrenkranz, Robert Hoffman, Claude Lamontagne, Carrie Spencer, Phil Winne and Carol Woodworth for their feedback.

References

Allport GW (1961) Pattern and growth in personality. Harcourt College Publishers, San Diego

Anderson JR, Milson R (1989) Human memory: an adaptive perspective. Psychol Rev 96(4): 703–719. doi:10.1037//0033-295X.96.4.703

Bandura A (1997) Self-efficacy: the exercise of control. W.H. Freeman, New York, p 604

Beaman CP, Williams TI (2010) Earworms (stuck song syndrome): towards a natural history of intrusive thoughts. Br J Psychol (London, England?: 1953) 101(Pt 4):637–653. doi:10.1348/000712609X479636

Beaudoin LP (1994) Goal processing in autonomous agents. Ph.D thesis, University of Birmingham, Birmingham, England

Beaudoin LP (2010a). Will the Apple tablet support or hinder users' cognitive fitness? Sharp-Brains. http://www.sharpbrains.com/blog/2010/01/26/will-the-apple-tablet-support-or-hinder-users'-cognitive-fitness. Accessed 26 January 2010

Beaudoin LP (2010b). Apple iPad thumbs-up: brain fitness [sic] value, and limitations [should have read "cognitive fitness"]. SharpBrains. http://www.sharpbrains.com/blog/2010/02/11/apple-ipad-thums-up-brain-fitness-value-and-limitations. Accessd 11 Feb 2010

Beaudoin LP (2011) Experts' productive learning from formal knowledge: motive generators and productive practice. In: Wyatt J (ed) A symposium in honour of Aaron Sloman: from animals to robots and back: reflections on hard problems in the study of cognition, Birmingham, UK, University of Birmingham, pp 139–172

Beaudoin LP (2013a) Cognitive productivity: the art and science of using knowledge to become profoundly effective. CogZest, Port Moody

Beaudoin LP (2013b) The possibility of super-somnolent mentation: a new information-processing approach to sleep-onset acceleration and insomnia exemplified by serial diverse imagining (MERP Report No. 2013-03). Meta-effectiveness research project, Faculty of Education, Simon Fraser University, p 40. http://summit.sfu.ca/item/12143

Beaudoin LP, Sloman A (1993) A study of motive processing and attention. In: Sloman A, Hogg D, Humphreys G, Partridge D, Ramsay A (eds) Prospects for artificial intelligence. IOS Press, Amsterdam, pp 229–238

Beaudoin LP, Winne P (2009) nStudy: an internet tool to support learning, collaboration and researching learning strategies. In: CELC-2009 Vancouver, BC. http://learningkit.sfu.ca/lucb/celc-2009-nstudy.pdf

Beck K (2002) Test driven development: by example. Addison-Wesley Professional, Boston, p. 240

Bereiter C (2002) Education and mind in the knowledge age. Routledge, London, p 544

Bereiter C, Scardamalia M (1993) Surpassing ourselves: an inquiry into the nature and implications of expertise. Open Court, Chicago, p 279

Boden M (1972) Purposive explanation in psychology. Harvard University Press, Cambridge

Carpenter SK, DeLosh EL (2006) Impoverished cue support enhances subsequent retention: support for the elaborative retrieval explanation of the testing effect. Mem Cogn 34(2):268–276

Craik FIM, Salthouse TA (2008) The handbook of aging and cognition. Psychology Press, New York, p 657

Csikszentmihalyi M (2008) Flow: the psychology of optimal experience. Harper Perennial Modern Classics, New York, p 336

Dennett DC (1987) The intentional stance. The MIT Press, Cambridge, p 400

Ericsson KA, Krampe RT, Tesch-Römer C (1993) The role of deliberate practice in the acquisition of expert performance. Psychol Rev 100(3):363–406. doi:10.1037//0033-295X.100.3.363

Ericsson K, Kintsch W (1995) Long-term working memory. Psychol Rev 102(2):211–245

Flippo RF, Caverly DC (eds) (2009) Handbook of college reading and study strategy research, 2nd ed. Routledge, New York, p 500

Fredrickson BL (1998) What good are positive emotions? Rev General Psychol 2(3):300–319

Frijda NH (1986) The emotions. Cambridge University Press, Cambridge, p 544

Gibson J (1994). Ref lists. Poplog Pop-11 documentation. http://wwwcgi.rdg.ac.uk:8081/cgi-bin/cgiwrap/wsi14/poplog/pop11/ref/lists. Accessed March 1994

Gladwell M (2000) The tipping point: how little things can make a big difference. Back Bay Books, Boston, p 279

Gottman JM, DeClaire J (2001) The relationship cure: a five-step guide for building better connections with family, friends, and lovers. Crown Publishers, New York, p 318

Gottman JM, Silver N (1999) The seven principles for making marriage work. Three Rivers Press, New York, p 288

Guadagnoli M (2009) Practice to learn, play to win. Ecademy Press, Cornwall, p 205

Haskell RE (2000) Transfer of learning: cognition and instruction. Academic Press, San Francisco

Jenkins R,(2002). Churchill: a biography. Plume, New York, p 1024

Karmiloff-Smith A (1995) Beyond modularity: a developmental perspective on cognitive science. In: Development. MIT Press, Cambridge, p 256

Karpicke JD, Blunt JR (2011) Retrieval practice produces more learning than elaborative studying with concept mapping. Science 331(6018):772–775

Karpicke Jeffrey D, Butler AC (2009) Metacognitive strategies in student learning: do students practise retrieval when they study on their own? Memory 17(4):471–479. doi:10.1080/09658210802647009

Kuo T-M, Hirshman E (1996) Investigations of the testing effect. Am J Psychol 109(3):451–464

Lakatos I (1980) The methodology of scientific research programmes: philosophical papers, vol 1. Cambridge University Press, Cambridge

Lamontagne C (2002) University teaching—a critical rationalist's reflexion (2001 award for excellence in teaching public address). Centre for University Teaching, Ottawa, pp 2–14

Latham GP, Locke EA (1991) Self-regulation through goal setting. Org Behav Human Decis Process 50(2):212–247. doi:10.1016/0749-5978(91)90021-K

Leach SK (2011) Personal communication.

McCarthy J (2008) The well-designed child. Artif Intell 172(18):2003–2014. doi:10.1016/j.artint.2008.10.001

McDaniel MA, Callender AA, (2008) Cognition, memory, and education. In: Roediger HL (ed) Learning and memory: a comprehensive reference, vol 2. Elsevier, Amsterdam, pp 235–244

Minsky M (2006) The emotion machine: commonsense thinking, artificial intelligence, and the future of the human mind. Simon & Schuster, New York, p 387

Minsky ML (1986) The society of mind. Simon & Schuster, New York, p 339

Mulcahy-Ernt PI, Caverly DC (2009) Strategic study-reading. In: Rona FF, Caverly DC (eds) Handbook of college reading and study strategy research. Lawrence Earlbaum Associates, Hillsdale, pp 177–198

Norman D, Bobrow DG (1979) Descriptions: an intermediate stage in memory retrieval. Cogn Psychol 11(1):107–123. doi:10.1016/0010-0285(79)90006-9

Oatley K, Johnson-Laird PN (1987) Towards a cognitive theory of emotions. Cogn Emot 1(1):29–50. Psychology. doi:10.1080/02699938708408362

Ortony A, Clore GL, Foss MA (1987) The referential structure of the affective lexicon. Cogn Sci Multidisc J 11(3):341–364

Ortony A, Clore GL, Collins A (1988) The cognitive structure of emotions. Cambridge University Press, Cambridge, p 207

Perkins D (1995) Outsmarting IQ: the emerging science of learnable intelligence. Free Press, New York, p 390

Perkins DN (1986) Knowledge as design. Laurence Earlbaum Associates, Hillsdale, p 231

Pinker S (2010) Mind over mass media. The New York Times (online). Retrieved from: http://www.nytimes.com/2010/06/11/opinion/11Pinker.html?_r=0

Popper KR, Eccles JC (1977) The self and its brain: an argument for interactionism. Routledge and Kegan Paul, London, p 597

Popper KR (1983) Realism and the aim of science: from the postscript to the logic of scientific discovery. In: Bartley-III WW (ed). Rowman and Littlefield, Totowa, p 420

Popper KR (1979) Objective knowledge. Oxford University Press, Oxford

Pressley M, Afflerbach P (1995) Verbal protocols of reading: the nature of constructively responsive reading. Laurence Earlbaum Associates, Hillsdale, p 157

Pyc MA (2010) Why is retrieval practice beneficial for memory? An evaluation of the mediator shift hypothesis, Kent State University

Reiss S (2000) Human individuality, happiness, and flow. Am Psychol 55(10):1161–1162. doi:10.1037//0003-066X.55

Roediger HL (2008) Relativity of remembering: why the laws of memory vanished. Ann Rev Psychol 59:225–254. doi:10.1146/annurev.psych.57.102904.190139

Roediger HL, Karpicke JD (2006b) The power of testing memory: basic research and implications for educational practice. Perspect Psychol Sci 1(3):181 (SAGE Publications)

Roediger HL, Finn B (2010) The pluses of getting it wrong. Sci Am Mind 21:38–41. doi:10.1038/scientificamericanmind0310-38

Ryle G (1949) The concept of mind. University of Chicago Press, Chicago, p 334 (2000 edition)

Schmidt BRA, Bjork RA (1992) New conceptualizations of practice: common principles in three paradigms suggest new concepts for training. Psychol Sci 3(4):207–218

Schopenhauer A (1841) Essays of Schopenhauer (Reading). (trans: Rudolf D). http://ebooks.adelaide.edu.au/s/schopenhauer/arthur/essays/

Schwartz DL, Bransford JD, Sears D (2005) Efficiency and innovtion in transfer multidisciplinary perspective. In: Mestre J (ed) Transfer of learning: research and perspectives. Information Age Publishing, Greenwich, pp 1–51

Selye H (1964) From dream to discovery: on being a scientist, 1st ed. McGraw Hill, New York, p 419

Simon HA (1967) Motivational and emotional controls of cognition. Psychol Rev 74(1):29–39

Sinclair N (2006) Mathematics and beauty: aesthetic approaches to teaching children. Teachers College Press, New York, p 196

Sloman A (1978) The computer revolution in philosophy: philosophy, science and models of mind. The philosophical review. Harvester Press, New York, p 300. doi:10.2307/2184449, http://www.cs.bham.ac.uk/research/projects/cogaff/crp/d:.

Sloman A (1984) The structure of the space of possible minds. In: Torrance S (ed) The mind and the machine: philosophical aspects of artificial intelligence. Ellis Horwood, Chichester, pp 35–42

Sloman A (1985) Why we need many knowledge representation formalisms. Expert Syst 1:1–17

Sloman A (1987) Motives, mechanisms, and emotions. Cogn Emot 1(3):217–233. doi:10.1080/02699938708408049

Sloman A (1993a) Prospects for AI as the general science of intelligence. In: Sloman A, Hogg D, Humphreys G, Partridge D, Ramsay A (eds) Prospects for artificial intelligence. IOS Press, Amsterdam, pp 1–10

Sloman A (1993b) The mind as a control system. In: Hookway C, Peterson D (eds) Philosophy and the cognitive sciences. Cambridge University Press, Cambridge, pp 69–110

Sloman A (1996) Towards a general theory of representations. In: Peterson DM (ed) Forms of representation: an interdisciplinary theme for cognitive science. Intellect, UK, pp 118–140

Sloman A (2008) The Cognition and Affect Project: Architectures, Architecture-schemas, and the New Science of Mind. Tech. rep., School of Computer Science, University of Birmingham. http://www.cs.bham.ac.uk/research/projects/cogaff/03.html#200307

Sloman A (2009) Architecture-based motivation versus reward-based motivation. http://www.cs.bham.ac.uk/research/projects/cogaff/misc/architecture-based-motivation.html

Sloman A (2010a) Phenomenal and access consciousness and the "hard" problem: a view from the designer stance. Int J Mach Conscious 02(01):117. doi:10.1142/S1793843010000424

Sloman A (2010b) Two notions contrasted: "logical geography" and "logical topography" variations on a theme by Gilbert Ryle: the logical topography of "logical geography". http://www.cs.bham.ac.uk/research/projects/cogaff/misc/logical-geography.html. Accessed 10 April 2011

Sloman A (2011) What's information, for an organism or intelligent machine? How can a machine or organism mean? In: Dodig-Crnkovic G, Burgin M (eds) Information and computation. World Scientific, Singapore, pp 1–32

Sloman A, Chrisley R (2010) Virtual machines and consciousness. J Conscious Stud 10(4):133–172

Stanovich KE (2009) What intelligence tests miss: the psychology of rational thought. Yale University Press, New Haven, p 328

Toplak ME, West RF, Stanovich KE (2012) Education for rational thought. In: Kirby JR, Lawson MJ (eds) Enhancing the quality of learning: dispositions, instruction, and learning processes. Cambridge University Press, Cambridge, pp 51–92

VanLehn K (1996) Cognitive skill acquisition. Ann Rev Psychol 47(1):513–539. doi:10.1146/annurev.psych.47.1.513

Vanlehn K, Arbor A, Jones RM (1991) What mediates the self-explanation effect? Knowledge gaps, schemas or analogies? In: 15th annual conference of the cognitive science society. Lawrence Erlbaum Associates, Hillsdale, pp 1034–1039

Vanlehn K, Arbor A, Jones RM (1993) Better learners use analogical problem solving sparingly. In: Proceedings of the tenth international conference on machine learning pp 338–345

Vlist E van der (2002) XML Schema. O'Reilly, Sebastopol, p 400

Wertheimer M (1959) Productive thinking. In: Michael W (ed). Harp & Brothers, New York, p 302 (Enlarged E)

Westar J (2009) Staying current in computer programming: the importance of informal learning and task discretion in maintaining job competence. In: Livingstone DW (ed) Education and jobs: exploring the gaps. University of Toronto Press, Toronto, pp 185–209

White AR (1964) Attention. Basil Blackwell, Oxford, p 134

White RW (1959) Motivation reconsidered: the concept of competence. Psychol Rev 66:297–333

Wiley J, Voss JF (1999) Constructing arguments from multiple sources: tasks that promote understanding and not just memory for text. J Educ Psychol 91(2):301–311. doi:10.1037//0022-0663. 91.2.301

Wilson J (1963) Thinking with concepts. Cambridge University Press, London, p 171

Winne PH (2001) Self-regulated learning viewed from models of information processing. In: Zimmerman BJ, Schunk DH (eds) Self-regulated learning and academic achievement: theoretical perspectives, 2nd edn. Lawrence Erlbaum, Mahwah, pp 153–189

Winne P (2006) How software technologies can improve research on learning and bolster school reform. Educ Psychol 41(1):5–17

Worden JW (1991) Grief counseling and grief therapy: a handbook for the mental health practitioner, 4th ed. Springer Publishing Company, New York, p 328

Wright IP, Sloman A, Beaudoin LP (1996) Towards a design-based analysis of emotional episodes. Philos Psychiatry Psychol 3(2):101–126

Chapter 13
From Cognitive Science to Data Mining: The First Intelligence Amplifier

Tom Khabaza

13.1 Introduction: Intelligence Amplifiers and Data Mining

The phrase 'Intelligence Amplification' (Ashby 1956; Licklider 1960; Engelbart 1962) refers to the idea that the products of Artificial Intelligence will be used initially, not to create fully intelligent machines, but to amplify or increase the power of human intelligence. Data mining (Berry and Linoff 1997; Helberg 2002) is one such intelligence amplifier; data mining algorithms form the core of a process which amplifies our ability to detect and act upon patterns in large quantities of data.

Whether data mining is really the first intelligence amplifier or not is open to debate; perhaps it is the first intelligence amplifier in widespread use. The purpose of this claim is to emphasise that data mining enhances our mental abilities in a way, which is much closer to the idea of intelligence amplification than most of the widespread use of IT.

13.2 Historical Background: Poplog, Clementine and CRISP-DM

During the 1980s, the Poplog AI programming environment (du Boulay et al. 1986) was developed at Sussex University under the leadership of Aaron Sloman, and was sold in the non-academic market by Systems Designers Ltd, which later became SD-Scicon. A management buyout from SD-Scicon in 1989 created Integral Solutions Ltd (ISL), whose core business was initially Poplog. ISL's initial product range included two machine learning modules based on decision trees and neural networks, and ISL's early business included a series of projects which applied machine learning to extract useful patterns from data, that is data mining projects

T. Khabaza (✉)
London, UK
e-mail: tom.khabaza@btinternet.com

J. L. Wyatt et al. (eds.), *From Animals to Robots and Back: Reflections on Hard Problems in the Study of Cognition*, Cognitive Systems Monographs 22, DOI: 10.1007/978-3-319-06614-1_13, © Springer International Publishing Switzerland 2014

(Fitzsimons et al. 1993). Based on the experience of these projects, Colin Shearer invented the Clementine data mining workbench (Khabaza and Shearer 1995).

Despite being the first practitioner to execute ISL's commercial data mining projects, I was initially sceptical about the prospects for data mining and the Clementine workbench. Clearly, the machine learning techniques used for data mining could not in themselves solve business problems of any significance; how then could data mining technology be of practical use?

The answer to this question, which emerged from successive projects, lay in the data mining process. Clementine had the (then unique) property of making data mining algorithms, at that time synonymous with machine learning algorithms, accessible to non-technologists. This meant that the process of understanding and preparing the data, applying the algorithms, and interpreting and using the results, could be executed by or in close collaboration with people whose primary knowledge was in the business domain (Shearer and Khabaza 1995). This in turn meant that business knowledge and understanding could be integrated closely with data mining technology in the process of business problem solving, without falling foul of the limitations of machine knowledge representation.

The design of Clementine, and the business-oriented data mining process which it enabled, were highly influential, and have shaped modern data mining practice and tools (e.g., Witten and Frank 2005; Sarma 2006; Silipo and Mazanetz 2012). This business-oriented process was later standardised in the data mining methodology CRISP-DM (Chapman et al. 1999).

13.3 Data Mining and CRISP-DM

Data mining is the use of business knowledge to create new knowledge, in natural or artificial form, by discovering and interpreting patterns in data. The term 'business' is used here to emphasise the use of data mining for practical purposes, but the definition would be equally correct if the word 'business' was replaced by 'domain'. At heart, data mining is a business process, and is used in a wide variety of applications, including customer analytics, fraud detection, risk management and law enforcement, and also in science and medicine.

The more recent term 'Predictive Analytics' usually refers to complete solutions in which data mining is embedded. Data mining is distinguished from other forms of data analysis by the use of 'data mining algorithms', also sometimes called 'predictive modelling algorithms'. 'Knowledge in artificial form' refers to the output of these algorithms, 'predictive models' or 'data mining models', which are used to increase information locally on the basis of generalisation, and are often embedded in Predictive Analytics solutions.

The industry standard data mining methodology is called CRISP-DM (which stands for CRoss-Industry Standard Process for Data Mining), and is depicted in Fig. 13.1 (from Chapman et al. 1999).

Fig. 13.1 CRISP-DM data mining methodology

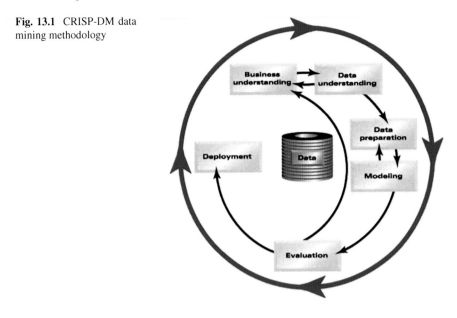

CRISP-DM was created by a research consortium, based on consultation with a wide circle of data mining practitioners. This consultation discovered that most practicing data miners had independently invented approximately the same process for successful data mining; CRISP-DM standardised the terminology and concepts of data mining, with many consequent benefits to communication, repeatability, training, education and consultancy.

13.4 Nine Laws of Data Mining

CRISP-DM provides an accurate picture of how data mining is carried out, but does not explain why the process has the form and properties that it does. Attempting to answer some nagging questions about data mining, I formulated the '9 Laws of Data Mining' (Khabaza 2010), and the explanations of these laws also help to explain the form of the data mining process.

These are not scientific laws in the usual sense of the phrase, because they are not based on a theory. Rather they are observations of the properties of data mining based on experience; their explanations are early steps towards a theory of data mining. The nine laws are listed below:

1. Business objectives are the origin of every data mining solution (Business Goals Law);
2. Business knowledge is central to every step of the data mining process (Business Knowledge Law);

3. Data preparation is more than half of every data mining process
   (Data Preparation Law);
4. The right model for a given application can only be discovered by experiment
   (or 'There is No Free Lunch for the Data Miner', NFL-DM);
5. There are always patterns (Watkins' Law);
6. Data mining amplifies perception in the business domain (Insight Law);
7. Prediction increases information locally by generalisation (Prediction Law);
8. The value of data mining results is not determined by the accuracy or stability
   of predictive models (Value Law);
9. All patterns are subject to change (Law of Change).

The rest of this section outlines the explanations for these laws.

First Law of Data Mining: 'Business Goals Law' *Business objectives are the origin of every data mining solution*

This defines the field of data mining, which is concerned with solving business problems and achieving business goals. Data mining is not primarily a technology; it is a process, which has one or more business objectives at its heart. Without a business objective (whether or not this is articulated), there is no data mining. Hence the maxim: 'Data Mining is a Business Process' (Khabaza 2002).

Second Law of Data Mining: 'Business Knowledge Law' *Business knowledge is central to every step of the data mining process*

This defines a crucial characteristic of the data mining process. A naive reading of CRISP-DM would see business knowledge used at the start of the process in defining goals, and at the end of the process in guiding deployment of results. This would be to miss a key property of the data mining process, that business knowledge has a central role in every step. For convenience, I use the CRISP-DM phases to illustrate:

- Business understanding must be based on business knowledge, and so must the mapping of business objectives to data mining goals. (This mapping is also based on data knowledge and data mining or analytical knowledge).
- Data understanding uses business knowledge to understand which data is related to the business problem, and how it is related.
- Data preparation means using business knowledge to shape the data so that the required business questions can be asked and answered. (For further detail see the third Law).
- Modelling means using data mining algorithms to create predictive models. The data miner applies business knowledge to formulate the models in terms of the data available, and also to interpret both the models and their behaviour in business terms. Understanding the business relevance of a model is necessary at all stages.
- Evaluation means understanding the business impact of using the models.
- Deployment means putting data mining results to work in a business process.

In summary, without business knowledge, not a single step of the data mining process can be effective; there are no 'purely technical' steps. Business knowledge guides the process towards useful results, and enables the recognition of those results that are useful. Data mining is an iterative process, with business knowledge at its core, driving continual improvement of results.

The reason behind this can be explained in terms of the 'chasm of representation' (an idea used by Alan Montgomery in data mining presentations of the 1990s). Montgomery pointed out that the business goals in data mining refer to the reality of the business, whereas investigation takes place at the level of data which is only a representation of that reality; there is a gap (or 'chasm') between what is represented in the data and what takes place in the real world. In data mining, business knowledge is used to bridge this gap; whatever is found in the data has significance only when it is interpreted using business knowledge, and anything missing from the data must be provided through business knowledge. Only business knowledge can bridge the gap, which is why it is central to every step of the data mining process.

Third Law of Data Mining: 'Data Preparation Law' *Data preparation is more than half of every data mining process*

It is a well-known maxim of data mining that most of the effort in a data mining project is spent in data acquisition and preparation. Informal estimates vary from 50 to 80%. Naive explanations might be summarised as 'data is difficult', and moves to automate various parts of data acquisition, data cleaning, data transformation and data preparation are often viewed as attempts to mitigate this 'problem'. While automation can be beneficial, it is a mistake to believe that it can remove the large proportion of effort which goes into data preparation; this would be to misunderstand the reasons why data preparation is required in data mining.

The purpose of data preparation is to put the data into a form in which the data mining question can be asked, and to make it easier for the analytical techniques (such as data mining algorithms) to answer it. Every change to the data of any sort (including cleaning, large and small transformations, and augmentation) means a change to the problem space which the analysis must explore. The reason that data preparation is important, and forms such a large proportion of data mining effort, is that the data miner is deliberately manipulating the problem space to make it easier for the available analytical techniques to find a solution.

There are two aspects to this 'problem space shaping'. The first is putting the data into a form in which it can be analysed at all—for example, most data mining algorithms require data in a single table, with one record per example. The data miner knows this as a general parameter of what the algorithm can do, and therefore puts the data into a suitable format. The second aspect is making the data more informative with respect to the business problem—for example, certain derived fields or aggregates may be relevant to the data mining question; the data miner knows this through business knowledge and data knowledge. By creating these fields and including them in the data for analysis, the data miner manipulates the search space to make it possible or easier for their preferred techniques to find a solution.

It is therefore essential that data preparation is informed in detail by business knowledge, data knowledge and data mining knowledge. These aspects of data preparation cannot be automated in any simple way.

The above also explains the otherwise paradoxical observation that even after all the data acquisition, cleaning and organisation that goes into creating a data warehouse, data preparation is still crucial to, and more than half of, the data mining process. Furthermore, even after a major data preparation stage, further data preparation is often required during the iterative process of building useful models, as shown in the CRISP-DM diagram.

Fourth Law of Data Mining: 'NFL-DM' *The right model for a given application can only be discovered by experiment or There is No Free Lunch for the Data Miner*

It is an axiom of machine learning that, if we knew enough about a problem space, we could choose or design an algorithm to find optimal solutions in that problem space with maximal efficiency. Arguments for the superiority of one algorithm over others in data mining rest on the idea that data mining problem spaces have one particular set of properties, or that these properties can be discovered by analysis and built into the algorithm. However, these views arise from the erroneous idea that, in data mining, the data miner formulates the problem and the algorithm finds the solution. In fact, the data miner both formulates the problem and finds the solution—the algorithm is merely a tool which the data miner uses to assist with certain steps in this process.

There are five factors which contribute to the necessity for experiment in finding data mining solutions:

(1) If the problem space were well-understood, the data mining process would not be needed—data mining is the process of searching for as yet unknown connections.
(2) For a given application, there is not only one problem space; different models may be used to solve different parts of the problem, and the way in which the problem is decomposed is itself often the result of data mining and not known before the process begins.
(3) The data miner manipulates, or 'shapes', the problem space by data preparation, so that the grounds for evaluating a model are constantly shifting.
(4) There is no technical measure of value for a predictive model (see eighth law).
(5) The business objective itself undergoes revision and development during the data mining process, so that the appropriate data mining goals may change completely.

This last point, the ongoing development of business objectives during data mining, is implied by CRISP-DM but is often missed. It is widely known that CRISP-DM is not a 'waterfall' process in which each phase is completed before the next begins. In fact, any CRISP-DM phase can continue throughout the project, and this is as true for Business Understanding as it is for any other phase. The business objective is not simply given at the start, it evolves throughout the process. This may be why some data miners are willing to start projects without a clear business objective—they know that business objectives are also a result of the process, and not a static-given.

Wolpert (1996) 'No Free Lunch' (NFL) theorem, as applied to machine learning, states that no one bias (as embodied in an algorithm) will be better than any other when averaged across all possible problems (datasets). This is because, if we consider all possible problems, their solutions are evenly distributed, so that an algorithm (or bias) which is advantageous for one subset will be disadvantageous for another. This is strikingly similar to what all data miners know, that no one algorithm is the right choice for every problem. Yet the problems or datasets tackled by data mining are anything but random, and most unlikely to be evenly distributed across the space of all possible problems—they represent a very biased sample, so why should the conclusions of NFL apply? The answer relates to the factors given above: because problem spaces are initially unknown, because multiple problem spaces may relate to each data mining goal, because problem spaces may be manipulated by data preparation, because models cannot be evaluated by technical means, and because the business problem itself may evolve. For all these reasons, data mining problem spaces are developed by the data mining process, and subject to constant change during the process, so that the conditions under which the algorithms operate mimic a random selection of datasets and Wopert's NFL theorem therefore applies. There is no free lunch for the data miner.

This describes the data mining process in general. However, there may well be cases where the ground is already 'well-trodden'—the business goals are stable, the data and its pre-processing are stable, an acceptable algorithm or algorithms and their role(s) in the solution have been discovered and settled upon. In these situations, some of the properties of the generic data mining process are lessened. Such stability is temporary, because both the relation of the data to the business (see 2nd law) and our understanding of the problem (see nineth law) will change. However, as long this stability lasts, the data miner's lunch may be free, or at least relatively inexpensive.

Fifth Law of Data Mining: 'Watkins' Law' *There are always patterns*

This law was first stated by David Watkins. We might expect that a proportion of data mining projects would fail because the patterns needed to solve the business problem are not present in the data, but this does not accord with the experience of data mining practitioners.

Previous explanations have suggested that this may be because:

- There is always something interesting to be found in a business-relevant dataset, so that even if the expected patterns were not found, something else useful would be found (this matches data miners' experience), and
- A data mining project would not be undertaken unless business experts expected that patterns would be present, and it should not be surprising that the experts are usually right.

However, Watkins formulated this in a simpler and more direct way: 'There are always patterns', and this provides a better match with the experience of data miners than either of the previous explanations. Watkins later amended this to mean that in data mining projects about customer relationships, there are always patterns connecting customers' previous behaviour with their future behaviour, and that these patterns can be used profitably ('Watkins' CRM Law'). However, data miners' experience is that this is not limited to CRM problems—there are always patterns in any data mining problem ('Watkins' General Law').

One possible explanation of Watkins' General Law runs as follows:

- The business objective of a data mining project defines the domain of interest, and this is reflected in the data mining goal.
- Data relevant to the business objective and consequent data mining goal is generated by processes within the domain.
- These processes are governed by rules, and the data that is generated by the processes reflects those rules.
- In these terms, the purpose of the data mining process is to reveal the domain rules by combining pattern-discovery technology (data mining algorithms) with the business knowledge required to interpret the results of the algorithms in terms of the domain.
- Data mining requires relevant data, that is data generated by the domain processes in question, which inevitably holds patterns from the rules which govern these processes.

To summarise this argument: there are always patterns because they are an inevitable by-product of the processes which produce the data. To find the patterns, start from the process or what you know of it—the business knowledge.

Discovery of these patterns also forms an iterative process with business knowledge; new patterns contribute to business knowledge, and business knowledge is the key component required to interpret the patterns. In this iterative process, data mining algorithms simply link business knowledge to patterns which cannot be observed with the naked eye.

If this explanation is correct, then Watkins' law is entirely general. There will always be patterns for every data mining problem in every domain unless there is no relevant data; this is guaranteed by the definition of relevance.

Sixth Law of Data Mining: 'Insight Law' *Data mining amplifies perception in the business domain*

How does data mining produce insight? This law approaches the heart of data mining—why it must be a business process and not a technical one. Business problems are solved by people, not by algorithms. The data miner and the business expert 'see' the solution to a problem, that is the pattern in the domain that allows the business objective to be achieved; data mining is, or assists as part of, a perceptual process. Data mining algorithms reveal patterns that are not normally visible to human perception. The data mining process integrates these algorithms with the normal human perceptual process, which is active in nature. Within the data mining process, the human problem solver interprets the results of data mining algorithms and integrates them into their business understanding, and thence into a business process. Section 13.5 below expands upon this theory.

Seventh Law of Data Mining: 'Prediction Law' *Prediction increases information locally by generalisation*

The term 'prediction' has become the accepted description of what data mining models do—we talk about 'predictive models' and 'predictive analytics' (Siegel 2013). This is because some of the most popular data mining models are often used to 'predict the most likely outcome' (as well as indicating how likely the outcome may be). This is the typical use of classification and regression models in data mining solutions.

However, other kinds of data mining models, such as clustering and association models, are also characterised as 'predictive'; this is a much looser sense of the term. A clustering model might be described as 'predicting' the group into which an individual falls, and an association model might be described as 'predicting' one or more attributes on the basis of those that are known.

Similarly we might analyse the use of the term 'predict' in different domains: a classification model might be said to predict customer behaviour—more properly we might say that it predicts which customers should be targeted in a certain way, even though not all the targeted individuals will behave in the 'predicted' manner. A fraud detection model might be said to predict whether individual transactions should be treated as high-risk, even though not all those so treated are in fact cases of fraud.

These broad uses of the term 'prediction' have led to the term 'predictive analytics' as an umbrella term for data mining and the application of its results in business solutions. Nevertheless we should remain aware that this is not the ordinary everyday meaning of 'prediction'—we cannot expect to predict the behaviour of a specific individual, or the outcome of a specific fraud investigation.

What, then, is 'prediction' in this sense? What do classification, regression, clustering and association algorithms and their resultant models have in common? The answer lies in 'scoring', that is the application of a predictive model to a new example. The model produces a prediction, or score, which is a new piece of information about the example. The available information about the example in question has been increased, locally, on the basis of the patterns found by the algorithm and embodied in the model, that is on the basis of generalisation or induction. It is important to remember that this new information is not 'data', in the sense of a 'given'; it is information only in the statistical sense.

Eighth Law of Data Mining: 'Value Law' *The value of data mining results is not determined by the accuracy or stability of predictive models*

Accuracy and stability are useful measures of how well a predictive model makes its predictions. Accuracy means how often the predictions are correct (where they are truly predictions) and stability means how much (or rather how little) the predictions would change if the data used to create the model were a different sample from the same population. Given the central role of the concept of prediction in data mining, the accuracy and stability of a predictive model might be expected to determine its value, but this is not the case.

The value of a predictive model arises in two ways:

- The model's predictions drive improved (more effective) action, and
- The model delivers insight (new knowledge) leading to improved strategy.

In the case of insight, accuracy is connected only loosely to the value of any new knowledge delivered. Some predictive capability may be necessary to convince us that the discovered patterns are real. However, a model which is incomprehensibly complex or totally opaque may be highly accurate in its predictions, yet deliver no useful insight, whereas a simpler and less accurate model may be much more useful for delivering insight.

The disconnect between accuracy and value in the case of improved action is less obvious, but still present, and can be highlighted by the question 'Is the model predicting the right thing, and for the right reasons?' In other words, the value of a model derives as much from of its fit to the business problem as it does from its predictive accuracy. For example, a customer attrition model might make highly accurate predictions, yet make its predictions too late for the business to act on them effectively. Alternatively, an accurate customer attrition model might drive effective action to retain customers, but only for the least profitable subset of customers. A high degree of accuracy does not enhance the value of these models when they have a poor fit to the business problem.

The same is true of model stability; although an interesting measure for predictive models, stability cannot be substituted for the ability of a model to provide business insight, or for its fit to the business problem. Neither can any other technical measure.

In summary, the value of a predictive model is not determined by any technical measure. Data miners should not focus on predictive accuracy, model stability, or any other technical metric for predictive models at the expense of business insight and business fit. In data mining, the only value is business value.

Nineth Law of Data Mining: 'Law of Change' *All patterns are subject to change*

The patterns discovered by data mining do not last forever. This is well-known in many applications of data mining, but the universality of this property and the reasons for it are less widely appreciated.

In marketing and CRM applications of data mining, it is well-understood that patterns of customer behaviour are subject to change over time. Fashions change, markets and competition change, and the economy changes as a whole; for all these reasons, predictive models become out-of-date and should be refreshed regularly or when they cease to predict accurately.

The same is true in risk and fraud-related applications of data mining. Patterns of fraud change with a changing environment and because criminals change their behaviour in order to stay ahead of crime prevention efforts. Fraud detection applications must therefore be designed to detect new, unknown types of fraud, just as they must deal with old and familiar ones.

Some kinds of data mining might be thought to find patterns which will not change over time—for example in scientific applications of data mining, do we not discover unchanging universal laws? Perhaps surprisingly, the answer is that even these patterns should be expected to change.

The reason is that patterns are not simply regularities which exist in the world and are reflected in the data—these regularities may indeed be static in some domains. Rather, the patterns discovered by data mining are part of a perceptual process, an active process in which data mining mediates between the world as described by the data and the understanding of the observer or business expert. Because our understanding continually develops and grows, so we should expect the patterns also to change. Tomorrow's data may look superficially similar, but it will have been collected by different means, for (perhaps subtly) different purposes, and have different semantics; the analysis process, because it is driven by business knowledge, will change as that knowledge changes. For all these reasons, the patterns will be different.

To express this briefly, all patterns are subject to change because they reflect not only a changing world but also our changing understanding.

13.4.1 Postscript to the Nine Laws

The nine Laws of Data Mining are simple truths about data mining; most are well-known to data miners, although some are expressed in an unfamiliar way (for example, the fifth, sixth and seventh laws). Most of the new ideas associated with the nine laws are in the explanations, which are an attempt to understand the reasons behind the well-known form of the data mining process.

Why should we care why the data mining process takes the form that it does? In addition to the appeal of knowledge and understanding, there is a practical reason to pursue these questions.

The data mining process came into being in the form that exists today because of technological developments—the widespread availability of machine learning algorithms, and the development of workbenches which integrated these algorithms with other techniques and made them accessible to users with a business-oriented outlook. Should we expect technological change to change the data mining process again? Eventually it must, but if we understand the reasons for the form of the process, then we can distinguish between technology which might change it and technology which cannot.

Several technological developments have been hailed as revolutions in predictive analytics, for example the advent of automated data preparation and model re-building, and the integration of business rules with predictive models in deployment frameworks. The nine Laws of Data Mining suggest, and their explanations demonstrate, that these developments will not change the nature of the process. The nine Laws can be used to evaluate such claims, in addition to their educational value for data miners.

13.5 From Intelligence to Perception

How and why does the data mining process produce new knowledge? The data mining process is essentially one of problem-solving; the business expert works out how to achieve an objective in the business domain. Business problems are solved by humans, not by algorithms, so how does data mining play a part in this?

The key issue addressed by data mining is that there may be useful information buried in data, where the required volume of data is too large for patterns to be seen unaided. (Watkins' Law states that such information is always present.) A conventional view of data mining would suggest that business goals are translated into data mining goals, then the algorithms are applied to the data, producing predictive models; these models are used to make predictions and help guide business decision-making in such a way as to help achieve the business goal. However, this view omits two crucial facts about data mining: one is the pervasive role of business knowledge (highlighted by the 2nd law), and the other is the production of insight, or new knowledge. It is on this second shortcoming that I will now focus.

While data mining may indeed produce predictive models to aid decision-making, both the models themselves and the process that produces them can also tell us new things about the business or domain. The process of understanding and preparing the data means examining the data in a great deal of detail, and new facts often emerge from this process; the data themselves have no intrinsic meaning, but when interpreted in the light of business knowledge the data often reveal important new information about the business, even before data mining algorithms are applied. When predictive models are produced, these will also often tell us important information about the business; this may be revealed by the behaviour of the model, or by the model itself, such as the readable rules in a decision-tree model, or by the relative importance of different input variables in unreadable models. Again this information has no intrinsic importance, but becomes important when interpreted in the light of business knowledge.

The production of insight at different stages of data mining can be illustrated by two examples. The first example concerns the production of insight as a by-product of preparing data for modelling. In this process, extensive data about a single entity (such as a customer) is often summarised to produce a 'behavioural segmentation' (for example, whether each customer makes high, medium or low use of a given product). Such behavioural segmentations often in their own right provide valuable insight into customer behaviour, in addition to being useful in predictive modelling. The second example concerns insights which arise from predictive models: one company discovered, from a readable predictive model, that certain aspects of their customer service policies where causing customers to cancel their contracts; the company then changed these policies, with a substantial benefit to customer retention.

It is characteristic of these processes that they take place in the business domain; every piece of data and every action has a business meaning. The data miner works, not in the realm of bits, bytes and algorithms, but in the domain of enquiry. The data mining process enables the data miner to see things in the domain which would not be visible unaided. We know that perception is an active, knowledge-based process. The data miner can 'see' things in the business domain because they 'know what they are looking at'.

My first hypothesis in this chapter is that data mining amplifies perception in the following way: data mining algorithms can detect patterns in data which are not visible to the naked eye, but the algorithms themselves have no domain knowledge. The business expert has the domain knowledge but cannot see the patterns unaided.

The data mining process (as described by CRISP-DM) enables the business expert to incorporate the pattern discovery capabilities of the algorithms into their own perceptual process. There is nothing mysterious about this; the process is mostly a codification of common sense, but it explains why data miners have the experience of 'seeing' things in the data. It is because data mining is like a perceptual process.

I have always wondered why machine learning algorithms (from the field of AI) seem to work better for data mining than those originating in the field of statistics. My second hypothesis in this chapter is that machine learning algorithms work well for data miners because they are designed to be part of a cognitive system. Machine learning techniques tend to be based on intuitively plausible models of knowledge. For the data miner, it matters little whether these models are correct descriptions of human cognition; what makes them useful for data miners is the plausible nature of the knowledge they create and the patterns they discover. This makes the algorithms easier to use as an extension of one's own cognition.

13.6 Conclusion: The Impact of Cognitive Science

A bird's-eye view of the activities of data miners in organisations would not immediately reveal anything to do with cognition. A data miner appears to (and does in fact) work in the domain of application; they would seem like marketeers, or fraud detection operatives, or police intelligence officers, or geneticists, or medics. They are exactly this, but also have their perceptual abilities, within their domain of operation, enhanced by the ability to see meaningful patterns in data. For data miners, data mining acts as an intelligence amplifier.

This kind of intelligence amplifier does not provide the expanded human intellect envisioned by Ashby (Asaro 2008); nevertheless, the expanded perceptual abilities of data miners can be used to make the world a better place (e.g., Van 2003; Piatetsky-Shapiro et al. 2003; Adderley and Musgrove 1999; McCue 2006; Chang and Shyue 2009).

If my second hypothesis is correct, then this ability of data mining to enhance the perception of domain workers is the result of the output of Cognitive Science research. By focusing on cognition, we have produced tools which can become part of cognition.

Acknowledgments I would like to thank Chris Thornton and David Watkins, who inspired the initial concepts behind this work, Chris Thornton again for his help in formulating NFL-DM, and also all those who have contributed to the LinkedIn discussion group '9 Laws of Data Mining', which has provided invaluable food for thought.

References

Adderley R, Musgrove PB (1999) Data mining at the West Midlands Police: a study of bogus official burglaries. BCS special group expert systems. Springer, London

Asaro PM (2008) From mechanisms of adaptation to intelligence amplifiers: the philosophy of W. Ross Ashby. In: Husbands P, Holland O, Wheeler M (eds) The mechanical mind in history. MIT Press, Cambridge

Ashby WR (1956) An introduction to cybernetics. Chapman and Hall, London

Berry MJA, Linoff G (1997) Data mining techniques: for marketing, sales and customer support. Wiley, New York

Chang C-J, Shyue S-W (2009) A study on the application of data mining to disadvantaged social classes in Taiwan's population census. Expert Syst Appl (Elsevier) 36:510–518

Chapman P, Clinton J, Kerber R, Khabaza T, Reinartz T, Shearer C (1999) CRISP-DM 1.0: step-by-step data mining guide. http://www.crisp-dm.org

du Boulay JBH, Khabaza T, Elsom-Cook M, Taylor J (1986) Poplog and the learner: an artificial intelligence environment used in education. In: Directory of computer training. Badegmore part Enterprises for Hoskyns Education

Engelbart DC (1962) Augmenting human intellect: a conceptual framework. Summary report AFOSR-3233. Stanford Research Institute, Menlo Park, CA

Fitzsimons M, Khabaza T, Shearer C (1993) The application of rule induction and neural networks for television audience prediction. In: Proceedings of ESOMAR/EMAC/AFM symposium on information based decision making in marketing, Paris, November 1993, pp 69–82

Helberg C (2002) Data mining with confidence, 2nd edn. SPSS, Chicago

Khabaza T, Shearer C (1995) Data mining with clementine. In: Proceedings of the IEE colloquium on knowledge discovery in databases, Digest No 1995/021(B), London, Feb 1995

Khabaza T (2002)Hard hats for data miners: myths and pitfalls of data mining. Data mining. WIT Press. Reprinted as DM review special report 2007

Khabaza T (2010) Nine laws of data mining. www.khabaza.com/9laws (also published as a discussion group on LinkedIn and on Twitter)

Licklider JCR (1960) Man–computer symbiosis. IRE Trans Human Fact Electron HFE 1:4–11

McCue C (2006) Data mining and predictive analysis: intelligence gathering and crime analysis. Butterworth-Heinemann, Burlington

Piatetsky-Shapiro G, Khabaza T, Ramaswamy S (2003) Capturing best practice for microarray gene expression analysis. In: Proceedings of the SIGKDD 2003, August 2003, Washington

Sarma KS (2006) Predictive modeling with SAS enterprise miner. SAS Institute Inc, Cary, NC

Shearer C, Khabaza T (1995) Data mining by data owners: presenting advanced technology to non-technologists through the clementine system. In: Intelligent data analysis '95. Baden-baden, Germany

Siegel E (2013) Predictive analytics. Wiley, New Jersey

Silipo R, Mazanetz MP (2012) The KNIME cookbook. KNIME Press, Zurich

Van J (2003) SPSS tools unravel secrets of disease. Chicago Tribune (2003) Retrieved from: http://articles.chicagotribune.com/2003-01-11/business/0301110153_1_spss-software-clementine-software-tool

Witten IH, Frank E (2005) Data mining: practical machine learning tools and techniques, 2nd edn. Morgan Kaufmann Elsevier, San Francisco

Wolpert D (1996) The lack of a priori distinctions between learning algorithms. Neural Comput 8:1341–1390

Chapter 14
Modelling User Linguistic Communicative Competences for Individual and Collaborative Learning

Timothy Read and Elena Bárcena

14.1 Introduction

The need to be able to understand and communicate in languages other than our own native tongue is an important skill in our modern networked society. As anyone who has dedicated any time and effort to learning a second language knows, progress is slow, and any break from the learning process causes such progress to be lost with alarming speed. Over the past decade, e-Learning has gone from being a minority learning approach used in some areas of distance education to becoming a key part of the educational process followed in the majority of traditional face-to-face learning institutions.[1] This transformation is argued to result from two changes in our modern society: first, to satisfy the need for *lifelong learning,* where mature students need/want to continue their studies and update their knowledge, competences and skills without the commitment of attending taught classes. Second, with the proliferation of broadband technology and the availability of low price computing hardware, students are used to having access to all sorts of Internet-based services, and generate a demand for them to fulfil part of their educational needs too. It is not, therefore, surprising that there is a lot of interest in using e-Learning systems for L2. However, these systems (or online educational platforms, virtual learning environments, etc.) provide nothing more than a means to shorten distances between students and their teachers, and suitable tools to enable the former to practise, interact and thereby learn. While this situation is relatively easy to achieve for small groups of students

---

[1] http://atlas.uned.es. Artificial Intelligence Techniques for Linguistic Applications (reference no FFI200806030).

T. Read (✉)
Departamento de Lenguajes y Sistemas Informáticos, UNED, Madrid, Spain
e-mail: tread@lsi.uned.es

E. Bárcena
Departamento de Filologías Extranjeras y sus Lingüísticas, UNED, Madrid, Spain
e-mail: mbarcena@flog.uned.es

J. L. Wyatt et al. (eds.), *From Animals to Robots and Back: Reflections on Hard Problems in the Study of Cognition*, Cognitive Systems Monographs 22, DOI: 10.1007/978-3-319-06614-1_14, © Springer International Publishing Switzerland 2014

and their corresponding teacher/ tutor, as the general demand for certain L2 learning increases, it becomes steadily more difficult to maintain such ratios. Another solution is required.[2]

Computers have been seen as tools which could play an important role in L2 learning since they first appeared (Levy 1997). However, typical Computer-Assisted Language Learning (henceforth, CALL) systems are quite limited and their impact on L2 learning has been small. To address this problem, Artificial Intelligence techniques have been added over the years to lead to a wide range of systems (henceforth, ICALL systems, e.g. Amaral and Meurers 2011; Wood 2008; Bailin 1995; Chanier 1994; Gamper and Knapp 2002; Holland et al. 1995), although with limited success. These systems typically focus on the most formal or organisational linguistic aspects, attempting to replace a human native speaker to correct erroneous student comprehension and/or production. Furthermore, where knowledge or domain modelling is undertaken, it is mostly done in a very ad hoc way, where the results of one system cannot be easily transferred to subsequent ones.

Most systems used for L2 learning present three problems: the oversimplification and reduction of the vastness and complexity of the learning domain to a few formal linguistic aspects (studied in closed and decontextualised activities), the lack of underlying pedagogic principles, and the complexity of automatic language parsing and speech recognition. To overcome these shortcomings, a theoretical framework which combines individual Cognitive Constructivism and collaborative Social Constructivism has been designed by the authors for implementation in ICALL systems (Bárcena and Read 2004, 2009, Read et al. 2002a, b, 2004, 2005, 2006). It simulates the way an experienced language teacher would interact with his/her students. The framework attempts to capture and model the relevant pedagogic, linguistic and technological elements for the effective development of L2 competence. One of the goals were that any ICALL system developed around this framework would structure the complex network of communicative language competences[3] (linguistic, pragmatic and sociolinguistic) and processes (reception, production and *interaction*) within the L2 learning process in a causal quantitative way, adapting such process to the progress made by a given student.

The authors argue that the essentially qualitative terms used in the *Common European Framework of Reference for Languages: Learning, Teaching, Assessment* (henceforth, CEFR; Council of Europe 2001). The CEFR need to be specified in a more quantitative way (so that they can be included in a computational model). Here, the modelling granularity is crucial, because when someone's global L2 competence is examined, what is really found is not a unitary *measure* of ability but heterogeneous degrees of capabilities that encompass the different modalities and functions, which go far beyond the traditional level of formality. Furthermore, while

---

[2] This research is part of the I-AGENT (Intelligent Adaptive Generic ENglish Tutor) project, and has been funded by the Spanish Ministry of Education (grant reference code: FFI2008-06030).

[3] Refer to the Common European Framework of Reference for Languages: Learning, Teaching, Assessment (Council of Europe 2001) for a precise interpretation of the concepts related to language use and learning used in this article.

the exact underlying cognitive and neuropsychological aspects involved in verbal communication acts are still largely a matter for academic debate, certain *external* aspects (general context, domain, spatial scenario, subject matter, status and mutual relationships, etc.) are directly observable and, therefore, it is possible to interpret them in quantitative terms that enable them to be modelled in ICALL systems. This is something that has not been tackled up to now.

14.2 Modelling Second Language Learning

Early research undertaken by the authors produced ad hoc models, which were needed for each different ICALL system. Such an approach lead to rather superficial models that became obsolete every time a new system was to be designed. Hence, it was evident that a deeper modelling activity was required to produce a conceptual L2 learning framework that could be used and extended for future applications. As such, the framework was developed (Read et al. 2002a, b, 2004, 2005, 2006) as a high-level abstraction of L2 learning and subsequently, the relation between the key elements argued to play a role in this process was specified (Bárcena and Read 2004, 2009). This relation is presented in Fig. 14.1 and its functioning can be summarised as follows. Initially, individual learning is undertaken through the performance of simple closed L2 activities, organised in a notional-functional way (van Ek and Alexander 1975; Wilkins 1976) and involving suitable tasks which include reading comprehension, pronunciation practice, new vocabulary learning, etc. Once there is evidence that prototypical conceptual learning starts to take place, collaboration becomes possible (for the same notion or concept) through the performance of more complex activities (typically involving several associated tasks). For collaboration to occur, the students working together must be capable of reaching mutual understanding. Such understanding requires communication in the L2 between the activity participants which, in turn, requires communicative strategies to be adopted (with the implicit intervention of what is known as *existential competence*, namely, the learner's personality features, motivations, attitudes, beliefs, etc., that influence his/her learning progress). The application of these strategies, therefore, permits collaboration to take place, reinforce previous individual learning, and trigger further individual study.

Over the last few years, there has been an extension to this relation. Since mobile devices are an ever more common tool for accessing online education systems, the framework has been extended to include the characteristics of ubiquitous learning as defined by Chen et al. (2002) and Curtis et al. (2002). Since these characteristics are neither cognitive nor collaborative, but have repercussions in both the student and group models, a third model, a ubiquity model is needed that fits into the framework as can be seen in Fig. 14.1.

The authors argue that a student's second language competence improves in relation to the way s/he moves between individual and collaborative learning processes. Here, student and group models are updated accordingly as progress is made, taking

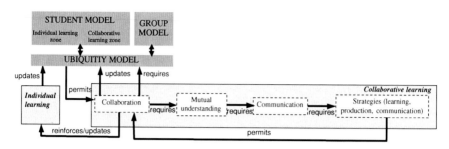

Fig. 14.1 Relation between individual, collaborative and ubiquitous learning

into account the aspects of information and device ubiquity present. The framework can, therefore, adapt the materials and activities to the current learning context. The underlying student and group models established and the ubiquity model needed to extend the framework are described subsequently, once the L2 domain is characterised and its models presented.

14.3 Characterising the L2 Domain

The fundamental problem when attempting to model the L2 domain is the complexity of natural language and the ways in which its use (and learning) can be specified. A significant breakthrough in this area came about with the publication of the CEFR (Council of Europe 2001). The CEFR offers, albeit in an essentially qualitative fashion, a way to structure the knowledge and skills required in second language learning. It is not surprising that it has been adopted in the majority of European L2 learning contexts, and even the prestigious L2 courses that different European countries are now certified in terms of the CEFR. It has been widely accepted (Morrow 2004) because it provides a notional-functional classification of language use and learning and it is the first general attempt to produce a taxonomy of the elements that intervene in language use and learning, enabling comparable syllabi to be created for all European languages. It captures the widespread functional and communicative perspective of human language and follows an action-oriented approach, which takes into account cognitive, volitional and emotional resources as well as the abilities specific to a learner both as an individual and as a social agent.

However, in order to use the CEFR as the linguistic reference for the L2 domain model, the authors have extracted eight fundamental concepts which enable a comprehensive and insightful quantitative representation to be established for the purpose of implementing I-CALL applications, namely:

1. *Language proficiency levels* The CEFR defines six language proficiency levels, as a general classification of a given student's L2 ability: from Breakthrough or

A1, the lowest level of generative language proficiency which can be identified, through to Mastery or C2.

2. *Communicative language competence* The most fundamental distinction made in the CEFR is between *communicative language competences* and *activities*. The former are the set of the learning resources (knowledge, skills, etc.), and the latter are what the learner can do with them.

3. *Competence descriptors* One of the key CEFR concepts used in the framework is that of *can-dos*, i.e. the language competence descriptors at a given common reference level (A1, A2, etc.). The CEFR offers a number of tables of illustrative descriptors which have been expanded and adapted for this domain model

4. *Contextualised language activities Prototypical activity* is used to refer generically to activities according to the direction of communication: *reception, production, interaction* and *mediation*. When a prototypical activity is seen in the context of one of the four spheres of reality that are distinguished in professional English, it is referred to as a *contextualised language activity*.

5. *Communicative language processes* There are defined in the CEFR to be the chain of neuropsychological and (and physiological) events involved in the reception and production of speech and writing. For the sake of computational modelling, these processes have been reduced to what is typically used in the literature as the 'four basic linguistic skills': reading, writing, listening, and speaking. While the communicative language activities are contextualised in the system according to the spheres of reality, these processes entail *modality*, i.e. their oral/written input/output nature.

6. *Domains/spheres of reality* The external context of language use in the CEFR can be seen to be divided into domains and situations. Four domains are distinguished: educational, occupational, public and personal. In this framework, however, the term *domain* has been substituted by *spheres of reality* because it is polysemic here since it has rather different meanings in sublanguage theory and also in AI. For the professional speaker, four spheres of reality have been identified in the framework: private (the sphere of intimate relationships), personal (the sphere of personal acquaintances), public (the sphere of general social relationships with other citizens), and occupational (the sphere of relationships in specialised working environments).

7. *Situations* The notions covered in the materials are also distributed across various *situations, locations* and *text types*, which constitute the external context of use of the language together with the spheres of reality. The CEFR distinguishes situations and, within these, locations, institutions, persons, objects, events, operations and texts. In order to simplify such descriptive parameters, the three categories with most impact on professional English were selected, namely locations (with attention to the type of related leisure/work activities), social roles (persons, with attention to mutual personal and/or work relationships) and text-types (texts).

8. *Texts* They provide samples of a particular type of notion and they embody vocabulary, grammatical forms and functions that are to be studied in relation to such a notion. Furthermore, language contents are organised in *topics*, and linked to the texts (there are between three to six topics for every text). It is essential to

note that texts in the 'CEFR sense' can in fact be written texts (e.g. e-mail, chats), audio sequences (e.g. radio recordings, phone conversations), video fragments (e.g. scenes both within and outside of the company) and images (e.g. plans, photographs).

As well as the CEFR-inspired concepts, two additional pedagogic features are included in the L2 domain models, namely: a spiral approach and the use of scaffolding. First, a common problem that any language teacher will be familiar with is that of a student who appears to have learnt a given language structure but later produces errors when trying to use it. This is an example of an internalisation problem, where short term learning is not transferred to long-term storage for future use. This can be a big problem for any L2 learner, since such 'knowledge holes' may eventually lead to a 'structural collapse' of the learning process. Hence, the *spiral approach* has been incorporated into the framework. Rather than following a structural approach to material presentation, where topics are tightly organised in sequences, the language difficulty is only partially graded and topics are reviewed 'in a spiral fashion', where the same topic is revisited now and again with an increasing level of complexity (Martin 1978). Second, *scaffolding* has been seen empirically as a learning facilitator. It is a didactic mechanism which provides support to the student when difficulties are detected by adding 'supportive devices', and gradually removes it as evidence of comprehension appears, a clear metaphor of the way in which buildings are actually scaffolded as they are built (McLoughlin and Marshall 2000; Bárcena and Read 2004).

14.4 Knowledge Meta-model and Models

Once a characterisation of the L2 domain has been produced for the framework, it is necessary to define the knowledge models (which have evolved over a period of years as different ICALL systems have been designed and explored). Emphasis is given to the reusability of information (one of the most persistent criticisms of AI systems is the large effort required to capture the knowledge domain which cannot be reused subsequently in other systems). As described in this section, the framework currently has one meta-model[4] and a set of knowledge models that can be split into two types: domain and student/ group models.

As has been noted above, the CEFR's essentially qualitative nature makes it impossible to directly use in any language learning system, so quantitative models that subsume and expand the CEFR have been developed. A central underlying structure for these models is the so-called **meta-model** that enables a representation of the necessary linguistic conceptual and functional knowledge, competences and skills in the L2 domain to be structured and interrelated. The meta-model takes the form of a 3D space that characterises language learning (illustrated in Fig. 14.2), in terms of

---

[4] A model that explains or underlies a set of other models.

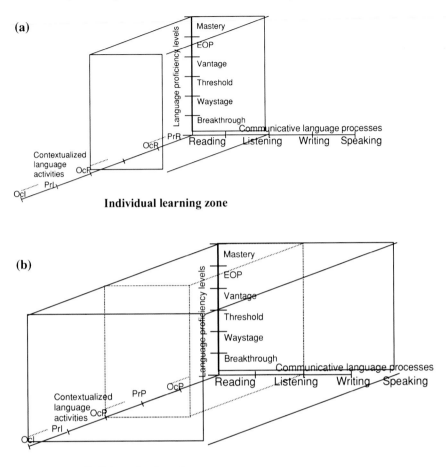

Fig. 14.2   The meta-model

the increasing *language proficiency levels* that are achieved by the units (dimension 1), the *communicative language processes* that are developed by the activation of *conceptual units* [5] (dimension 2), and the *contextualised language activities* that can be performed by such activation (dimension 3). In such a fine-grained activity-based characterization of communicative language competence assessment, a distinction is made according to the learning phase. This reflects the degree of attention being applied by the student on the application of a particular conceptual unit, which is a determining factor to assess the level of assimilation of such unit. This meta-model is divided into two parts: the individual and collaborative learning zones (shown as

---

[5] A fundamental competence item, formed by the intersection of the three dimensions. These items can be seen represented graphically by the dots in Fig. 3.7.

A and B). The intersection of each dimension does not define a small set of values that portrays learning, as if it were a 'pure' stereotype student model (Kay 2000), but contributes to an overall communicative language competence state, following a multidimensional stereotype model. This space is initially empty and then gradually populated by points corresponding to conceptual units, which are related across the learning dimensions, as the students learn and apply them in different activities and processes.

This meta-model provides a fine-grained representation of L2 learning, since at any given time, a student is not completely 'at a particular proficiency level', in the sense of having all the points within the 3D space at the same level, but has the points distributed over adjacent levels, depending on past experience and the state of consolidation of the conceptual units in different communicative language processes and contextualised language activities. The uniform progression within the space ensures that knowledge holes or gaps do not appear and therefore prevents structural collapse of a student's communicative language competence (Read et al. 2002a, b). It also ensures balanced learning between the different dimensions that make it up: communicative language processes and contextualised language activities, e.g. a learner of English who has very good vocabulary (because he reads endlessly in this language) will not be left to increase his lexical repository and reading at the expense of neglecting his listening or speaking skills, so that he progresses effectively in his learning.

Now that the underlying meta-model has been detailed, it is possible to move on to detail the actual knowledge models that use it in the framework. The L2 domain is represented in two models: one represents the linguistic and didactic information, and the other, the cognitive and collaborative information about the students and how they work to learn. The problem domain is made up of four models: conceptual, linguistic knowledge, collaborative template and content. They can be described as follows:

- *M1*: the *conceptual model* defines the knowledge domain for learning professional English. In this model, the can-dos are classified for the activities within the spheres of reality and the communicative language processes at the different language proficiency levels.
- *M2*: the *linguistic knowledge model* relates each can-do to the grammar, semantics, and discourse topics that a student needs to learn in order to 'be competent' in the particular can-do.
- *M3*: the *collaborative template model* defines the structure of the activities and tasks that the students undertake in groups, together with references to the tools and resources they use.
- *M4*: the *content model* contains the actual linguistic (and related) knowledge that the students learn when working with the system. Different types of didactic materials are stored to offer students a wide range of learning options and practice depending, among other things, upon previous study sessions and results.

The **student model** (shown in Fig. 14.3) represents the state of communicative language competence together with the profile and history of a student's individual and collaborative activities.

This model can be seen to be divided into four parts:

1. *The personal details of the student* This information defines the way in which a student prefers to work with the system: in a system driven fashion or as a mixed initiative process. In the former, the system leads the student through the learning process. In the latter, the student is granted a certain degree of study flexibility in terms of the selection and type of materials.
2. *The individual learning activities undertaken with the system* This is a log that prevents unnecessary repetition by recording the list of activities that a student has done. This information is important not only for current system functioning, but also for future analyses of effective ways (or 'study paths') in which students work.
3. *The collaborative activities undertaken by a student* This takes the form of references to instances of group models, which record the roles, tasks, etc., undertaken, and the results obtained (as will be seen below, groups are adaptively formed and only last for the life of an activity).
4. *The English communicative language competence characterisation* This is an instantiation of the information contained in the 3D meta-model.

A further model is required both to structure and to coordinate the way in which the students work together: the **group model**. This model stores collaborative data and represents the details of the set of students working on particular tasks within an activity, detailing interactions, mistakes, etc. This model (shown in Fig. 14.4) is a single register that characterises the collaborative activity of the group. The data in the model comes partially from the questions a student-monitor (a student in the group with a supervising role) has to answer when the group is evaluated, and partially from the activity log.

The information contained in this model is required since adaptive group formation requires a history of previous student group activities when assigning roles and membership. Furthermore, the log of these group models permits the learning process undertaken by each student to be analysed so that the overall didactic properties of the framework can be evaluated. The student and group models have different functional roles in the system. A student model is instantiated when a student enters the system for the first time, and exists until the student is removed (administratively speaking) from the system. A group model in contrast, once formed, is active and available only during a particular collaborative activity. Once the activity has finished, the group is dispersed, the group data is logged as part of the 'history of the learning process', and the group model deactivated.

Finally, a **ubiquity model** was developed that incorporated the content, device and language learning context characteristics that are necessary for representing the various degrees of ubiquity which a second language learner might encounter (as can be seen in Fig. 14.5). It is important that the content is reliable, accurate and

STUDENT MODEL	
PERSONAL INFORMATION	
Name	User Id Login Password
Initiative:	System / Mixed
Objectives:	Improve language proficiency / Fill a knowledge gap / Practise a genre
Degree of theoretical explanation:	Standard / Minimum
Other preferences and/or restrictions	
INDIVIDUAL ACTIVITIES	
Topics studied:	Theory Examples Activities
COLLABORATIVE ACTIVITIES	
References to group models in which the student has worked / is working.	
ENGLISH LANGUAGE PROFICIENCY LEARNING CHARACTERISATION	
Conceptual units:	Notions, Competence descriptors, Topics, Activities (tasks)
Language proficiency level:	Breakthrough, Waystage, Threshold
Linguistic level:	Grammar, Semantics, Discourse
Communicative language processes:	Reading, Listening, Writing, Speaking
Contextualized language activities:	PrR … OcI
Learning phase:	Attentional learning / Non-attentive application

Fig. 14.3 The information represented in the student model

GROUP MODEL	
Group	Id Group -
Activity	Id Activity
GROUP INDICATORS	**(values)**
State	Not Initiated /Initiated/Working/Impasse/Out of Time/ Finished
Role Assignment	Inadequate/Compensated/Good/Very Good
Participation	Low/Adequate/Active/Very Active
Effectiveness	Poor/Adequate/Good/Very Good
GROUP MEMBER INDICATORS	**(values; one for each group member)**
Student ID	Id Member
SM	Reference to the student model of this member
Role	Role in the activity
Role development	Poorly /Adequate/Well /Very Well
Interaction Level	Understanding/Communication/Conversation/Discussion/Negotiation
Participation	Low/Adequate/Active/Very Active
Decision Making	Very Bad/Needs Improvement/Adequate/Good/Very Good
Promotes discussion	Passive/Conformed/Active/Very Active
Critical thinking	Needs improvement/Adequate/Outstanding
Initiative	Passive/Low/Adequate/Very Good

Fig. 14.4 The information represented in the group model

structured for users at different knowledge and expertise levels. Content must be fully understandable and navigable, ideally not depend upon one particular type of software or hardware, and be suitable for different types of devices (e.g. screen resolutions, colours, sizes, etc.). In essence, it must be completely user-centred, and not

UBIQUITY MODEL	
CONTENT INFORMATION	
Information structure:	None / Type
Interoperability level:	Very Low / Low / Medium / High / Very high
Standardisation of resources:	None / LOM / IMS-LD / IMS-CC / SCORM
Need of adaptation (usage of ICT Assistive Technologies):	Any / Visual / Auditive / Cognitive
Interaction level:	Expositive / Active / Interactive
Data persistency and redundancy:	None / Type
DEVICE INFORMATION	
Localisation:	Zone
Connection type:	MOBILE (Wireless, GPRS, ...) / DESKTOP (LAN, ...) / etc.
Device type:	PDA / Mobile phone / etc.
Multimedia capabilities:	None / Type
Multimedia formats:	Flash / MPEG
Connection reliability:	Yes / Permanent sessions / Not always / None
Interoperability level:	None / Fully interoperable
COMMUNICATIVE LANGUAGE LEARNING CONTEXT	
Communicative language competences descriptors:	Cognitive competences / Language learning strategies / Contextualised language activities / Communicative language processes
Scaffolding mechanisms:	Context dependant help / Additional activities following location / Peer communication / Hints / etc.
Peer monitoring and diagnosis:	Possible types / Temporal restrictions / Importance / etc.
Availability level:	Location type / Activity type / Available time / Peer contact possible / etc.

Fig. 14.5 The information represented in the ubiquity model

device-centred. Regarding the device, the potential for providing different degrees of ubiquity in a learning environment requires the student to be provided with the content or means to develop certain activities at any place, any time, and of course, undertake them with no problems. Communication (student–student, student–teacher) should always be set up so that a system or platform works correctly and allows the concurrency of several users for collaborative purposes. Therefore, connectivity issues must be assured so that they do not restrict or disrupt learning, because this could have a negative effect on student progression. Even though the available technical infrastructure is a key factor, this term cannot be included directly as an indicator inside the model, since it is device aspects and content interoperability issues that measure the efficiency of the technical infrastructure.

As well as information regarding the content and device, this model represents the degree of ubiquity present for given language learning activities, devices and situations in terms of a communicative language learning environment. This environment specifies the constraints which enable language activities to be carried out following the CEFR, and has been refined by the authors from previous work (Read et al. 2006). It defines the specific way in which a particular handheld device can be used in a given real world context to undertake language learning using this framework.

14.5 Framework Applications and Conclusions

The framework presented in this article and its knowledge models have been iteratively refined by use in different ICALL applications. In each case, test study results have shown improved learning (with the relevant system) when compared to individual study in e-Learning groups. Specifically, four systems are representative of the main milestones in this research: I-PETER I, I-PETER II, COPPER and I-AGENT. I-PETER I (Intelligent Personalised English Tutoring EnviRonment; Read et al. 2002a) was the first system to use this framework in its current version at the time. This system contains no collaborative learning and enables error diagnosis of individual learning activities to be undertaken using a Bayesian network, to reflect how teachers actually undertake this type of process in the classroom. The results of this diagnosis process enable a finer-grained control of material selection than is normally possible, giving rise to a course structure that is continuously adapted to individual student needs. I-PETER II takes off where I-PETER I finishes (Read et al. 2004), using the same Bayesian diagnosis process for a larger domain (a Business English course; 56 networks were used) for individual learning together with peer and tutor corrected collaborative activities. For the first time, the activities were structured to combine differential knowledge use (mechanical reproduction + non-attentive application).

COPPER (Collaborative Oral and written language adaPtive Production EnviRonment; Read et al. 2005) combines individual and collaborative learning. This is the first system to use the full framework as detailed in this article. Language use is conceptualised in this system as one of several cognitive competences that are mobilised and modified when individuals communicate. High granularity expert-centric Bayesian networks with multidimensional stereotypes are used to update student's activities semi-automatically. An adaptive group formation algorithm dynamically generates communicative groups based upon the linguistic capabilities of available students, and a collection of collaborative activity templates. The results of a student's activity within a group are evaluated by a student monitor, with more advanced linguistic competences, thereby sidestepping the difficulties present when using natural language processing techniques to automatically analyse non-restricted linguistic production. Students therefore initially work individually on certain linguistic concepts, and subsequently participate in authentic collaborative communicative activities, where their linguistic competences can develop approximately as they would in 'real foreign language immersion experiences'. The work produced in two COPPER test groups in a final task undertaken in a pilot study, was both quantitatively and qualitatively superior to that in the first one: it was more intelligible, more idiomatic, had less mistakes, and was more adequate for the communicative scenarios that had been created for the activities. Such an improvement was less noticeable in the control group.

Currently, the system I-AGENT (Intelligent Adaptive Generic ENglish Tutor; Bárcena and Read 2009) is being finished, which is the first system to substitute the use of Bayesian networks for reasoning techniques taken from the field of Web Semantics, although the framework presented here is used in its entirety. It is also different to previous systems in that it integrates collaborative online work via Moodle and face-to-face classroom lessons (i.e. it uses a blended learning approach).

In this article, the models that make up an innovative framework for use in ICALL systems have been presented. It has been argued that such a specification is needed to facilitate the design and development of general purpose ICALL systems that can be used by L2 teachers and students. Future work is needed to focus on how to incorporate the modelling of neuropsychological aspects of language use and learning contained in the CEFR (developing concepts such as existential competence, the mental contexts of the interlocutors, intentions, lines of thought, expectations, states of mind, memory effects, etc.), and their incorporation in both the individual and collaborative parts of the framework. For example, the study of the optimum way in which roles can be changed within a community of students (as a result of these aspects) when undertaking different activities in new groups may also prove to be a fruitful line for future research in the field of collaborative learning.

Regarding subsequent evaluation of the framework, a large-scale test will be necessary to study how students actually improve their overall communicative professional English competence. A much larger experiment (with more students at different communicative language competence states, an extensive sample of activities and groups, etc.) will facilitate a more controlled and contextualised study of the system. In order to evaluate each of its features, parallel experimental groups will be needed that work with controlled versions of the system in order to allow a given feature to be tested in isolation. This work is important because, as well as enabling linguistic improvement to be evaluated, it will also permit the robustness and scalability of the computational implementation to be tested.

References

Amaral L, Meurers D (2011) On using intelligent computer-assisted language learning in real-life foreign language teaching and learning. In: ReCALL, vol 23, no 1. Cambridge University Press, Cambridge, pp 4–24

Bailin A (1995) AI and language learning: theory and evaluation. In: Holland VM, Kaplan JD, Sams MR (eds) Intelligent language tutors: theory shaping technology. Lawrence Erlbaum Associates, Mahwah, pp 327–343

Bárcena E, Read T (2004) The role of scaffolding in a learner-centered tutoring system for business english at a distance. In: Proceedings of the 3rd EDEN research workshop, Universität Oldenburg, pp 176–182

Bárcena E, Read T (2009) The integration of ICALL-based collaboration with the english classroom. In: To be published in proceedings of EuroCALL 2

Chanier T (ed) (1994) Language learning. Spec. Issue J. Artif. Intell. Educ 5(4)

Chen YS, Kao , Sheu JP, Chiang CY (2002) A mobile scaffolding-aid-based bird -watching learning system. In: Proceedings of IEEE international workshop on wireless and mobile technologies in education (WMTE'02), Växjö (Sweden). IEEE Computer Society Press, pp 15–22

Council of Europe (2001) Common European framework of reference for languages: learning, teaching, sssessment. Cambridge University Press, Cambridge

Curtis M, Luchini K, Bobrowsky W, Quintana C, Soloway E (2002) Handheld use in K-12: a descriptive account. In: Proceedings of IEEE international workshop on wireless and mobile technologies in education (WMTE'02). IEEE Computer Society Press, Växjö, Sweden, pp 23–30

Gamper J, Knapp J (2002) A review of intelligent CALL systems. Comput Assist Lang Learn 15(4):329–342

Holland VM, Kaplan JD, Sams MR (eds) (1995) Intelligent language tutors, theory shaping technology. Lawrence Erlbaum Associates, Mahwah

Kay J (2000) Stereotypes, student models and scrutability. In: Gauthier G, Frasson C, VanLehn K (eds) Intelligent tutoring systems. ITS2000 (LNCS 1839). Springer, Heidelberg, pp 19–30

Levy M (1997) Computer assisted language learning: context and conceptualization. Oxford University Press, Oxford

Martin MA (1978) The application of spiraling to the teaching of grammar. TESOL Quart 12(2):151–161

McLoughlin C, Marshall L (2000) Scaffolding: a model for learner support in an online teaching environment. In: Proceedings of the teaching and learning, Forum (2000)

Morrow K (2004) Insights from the common European framework. Oxford University Press, Oxford

Read T, Bárcena E, Barros B, Verdejo MF (2002a) I-PETER: modelling personalised diagnosis and material selection for an on-line english course. In: Garijo F, Riquelme J, Toro M (eds) Advances in artificial intelligence—Iberamia 2002 (LNAI 2527). Springer, Berlin, pp 734–744

Read T, Bárcena E, Barros B, Verdejo MF (2002b) Adaptive modelling of student diagnosis and material selection for on-line language learning. J Intell Fuzzy Syst 12(3/4):135–150

Read T, Bárcena E, Pancorbo J (2004) Adaptive tutoring architectures for english distance learning. In: Proceedings of the international conference on education. (Innovation. Technology and Research in Education), Bilbao, pp 142–146

Read T, Bárcena E, Barros B, Varela R, Pancorbo J (2005) COPPER: modeling user linguistic production competence in an adaptive collaborative environment. In: Ardissono L, Brna P, Mitrovic A (eds) 10th international conference on user modelling, Edinburgh, (LNAI 3538). Springer, Berlin, pp 144–153

Read T, Barros B, Bárcena E, Pancorbo J (2006) Coalescing individual and collaborative learning to model user linguistic competences. In: Gaudioso H, Soller A, Vassileva J (eds) User modeling and user-adapted interaction. Special issue on user modeling to support groups, communities and collaboration, vol 16, no 3/4, pp 349–376

Read T, Bárcena E, Rodrigo C (2010) Modelling ubiquity for second language learning. Int J Mob Learn Organ 4(2):130–149

van Ek JA, Alexander LG (1975) Threshold level english. Pergamon Press, Oxford

Wilkins DA (1976) Notional syllabuses. Oxford University Press, Oxford

Wood P (2008) Developing ICALL tools using GATE. In: Computer assisted language learning, vol 21, no 4

Chapter 15
Loop-Closing Semantics

Ian Wright

> *'In principle, if you want to explain or understand anything in human behavior, you are always dealing with total circuits, completed circuits. This is the elementary cybernetic thought.'*

Gregory Bateson (1999)

15.1 Semantic Properties

I believe the temperature of the room is 20 °C. Clearly, my belief is not the same thing as the temperature of the room. My belief *refers* to something else—it has semantic content.

My belief picks out some isolated feature or features of the world, while excluding all others. For example, my belief does not refer to air pressure. So the semantic content is *focused* and often univocal.

And I may be mistaken. My beliefs have truth conditions; they are defeasible. For example, it just so happens that the room temperature is 15 °C. My belief has a *semantic value*, which here is "false".

Can these semantic properties—reference, focus, and semantic value—be reductively explained in terms of a theory that does not presuppose semantic properties? Can intentional phenomena be reduced to known nonintentional phenomena and therefore naturalized?

In this chapter, I address this question with reference to simple artificial artifacts, especially thermometers and thermostats. These simple devices are *prima facie* unpromising candidates for admission into the class of things that (nonderivatively)

I. Wright (✉)
Economics, Faculty of Social Sciences, Open University, Milton Keynes, UK
e-mail: wrighti@acm.org

J. L. Wyatt et al. (eds.), *From Animals to Robots and Back: Reflections on Hard Problems in the Study of Cognition*, Cognitive Systems Monographs 22, DOI: 10.1007/978-3-319-06614-1_15, © Springer International Publishing Switzerland 2014

possess intentional properties. But unlike minds they possess no "black box" mysteries that force us to conjecture about how they really work. Since they are so simple they can be understood in every detail. Hence, if we demonstrate that some simple artifacts in fact possess intentional properties, independent of our practices, then we will have a clear understanding of exactly how (some) intentional properties are reducible to nonintentional properties.

Although I make extensive use of simple examples, the ultimate explanatory target is, of course, cognition in general—whether simple, complex, natural, or artificial.

First, I consider a "straw man" theory of semantic properties, and point out its problems, in order to introduce the explanatory requirements that any account of semantic properties must satisfy.

15.2 A Crude Causal Theory

Consider a thermometer that indicates temperature by the height of its mercury column, which is suspended in a capillary tube. As the local temperature rises, the mercury expands; when the temperature falls, it contracts. The column height is calibrated against a standard scale (e.g., °C), which we read off.

A necessary condition for a thermometer to measure temperature is the lawful covariation of some part of it with temperature. For example, the design of the thermometer exploits the natural law of thermal expansion. The height of the mercury column is an "information-bearing sub-state" (Sloman 1994b), or "representational vehicle", that reliably indicates the local temperature. We might propose, then, that the substate represents temperature in virtue of this lawful covariation.

This proposal forms the theoretical core of a family of "information semantic" Fodor (1990, Ch. 3) theories of content. Typical examples are (Dretske 1981), Fodor (1990), Jacob (1997) and Barwise and Seligman (1997) (although, I should stress, each of these examples considerably extends and alters this core and is not reducible to it). Information semantic theories maintain, in one form or another, this initial premise:

(IS) Information-bearing sub-state X refers to Y if Y reliably causes the state of X.

For example, the height of the mercury column refers to temperature because temperature reliably causes its height. The precise specification of "reliable cause" differs between theories; it may be expressed in terms of conditional probabilities, "nomic regularities" (laws), or sets of counterfactual dispositions, etc.

Some mental representations are not obviously caused by what they refer to (e.g., imagining flying pigs) but information-bearing sub-state (IS) makes no claim to be a complete theory of representation.

Let us now examine the problems with this "crude causal theory" (Fodor 1989, Ch. 4).

15.3 Conjunction Problems

Consider this complicated setup. Submerge a heating element in a container filled with a liquid. Float a small thermokinetic engine on the liquid's surface. Connect the engine to a pulley system that opens or closes an aperture connected to an independent source of hot air. Place a thermometer before the aperture. Now turn on the heating element. As the liquid's temperature rises, the heat induces rotary motion in the engine which opens the aperture blowing hot air onto the thermometer. (In general, we can imagine arbitrary "Rube Goldberg" machines in-between the object whose temperature is to be measured and the thermometer.)

In this specific setup, the state of the thermometer's information-bearing substate is (at least) reliably caused by the output of the heating element, the liquid's temperature, the speed of the engine, the size of the aperture, and the amount of hot air flowing over the thermometer. According to IS the height of the mercury column represents the output of the heating element *and* the liquid's temperature *and* an engine speed *and* the size of an aperture *and* the flow of hot air, or any subset of this conjunction.

The point is this: a reliable causal chain may be complex and therefore support multiple "upstream" candidates for the semantic content of any "downstream" information-bearing substate in that chain. IS picks out the *conjunction* of all the antecedent covarying features as the semantic content of the thermometer. The thermometer's information-bearing substate is therefore semantically indeterminate: we have a "conjunction problem". So IS has difficulty in explaining semantic focus since it provides no basis—independent of our intentional practices—for claiming that a thermometer represents (only) temperature.

Rube Goldberg machines are highly contrived. Perhaps we should exclude such unusual situations. In normal circumstances, most of the time, temperature is the only feature of the world that reliably causes the height of the mercury column. Can we therefore reject the conjunction problem on these grounds?

No. The premise is false. Temperature is not the only feature that reliably causes the height of a mercury column. For example, a thermometer that measures air temperature also normally measures air pressure (Jacob 1997, p. 103) since temperature and pressure are properties that often covary (Gay-Lussac's law). And a thermometer that measures the temperature of a liquid also normally measures the rate of evaporation of molecules from the surface of the liquid.

Rube Goldberg machines, therefore, are not required to expose the conjunction problem. In fact, highly contrived, and therefore unusual, experimental situations—such as a thermometer measuring the temperature of the air inside a balloon or the temperature of a completely sealed liquid—are necessary to exclude air pressure and evaporation as covarying properties. We normally ignore these covarying features when thinking of the function of a thermometer but a reductive theory of semantic properties cannot.

Perhaps we can exclude distal features of the causal chain—color, engine speed, aperture size, etc.—by restricting IS to the *proximal* feature, or final cause, of the height of the mercury column? The final cause, in our example, is hot air blown across the thermometer. Such a modified theory would pick out temperature (and not color, engine speed, or aperture size).

But this restriction fails to generalize since the final cause of a thermometer's reading need not be an instance of the natural kind "temperature". For example, infrared thermometers, widely available, measure temperature at a distance by focusing infrared light from the measured object to an internal detector. The proximal feature, in this case, is not temperature but light. (Furthermore, it is nonsensical to to claim that when I read a nearby thermometer, and form the belief that the room temperature is 20 °C, then my belief, in fact, refers to the proximal projection of light on the surface of my retina.) The distinction between distal and proximal features does not solve the conjunction problem.

15.4 Disjunction Problems

Beliefs can be in error. They have semantic values, such as "true" of "false". So any reductive theory of semantic content needs to explain how a representation can misrepresent, or fail to refer.

Place a mercury thermometer in a bowl of dry ice that is below −40 °C. The mercury freezes solid. In consequence, the thermometer registers the wrong temperature, say −35 °C, and the semantic value of its information-bearing substate is false.

IS claims that the height of the mercury column represents temperature because temperature reliably causes it. At very low temperatures, mercury freezes solid and disrupts this causal relationship. At first glance, therefore, IS appears to successfully explain misrepresentation.

But IS appears to work only because we (almost reflexively) foreground circumstances in which the thermometer functions as intended from circumstances in which it does not. But our beliefs about "normal circumstances" do not figure in IS. In fact, the height of *solid* mercury is also reliably caused by temperature with the difference that the coefficient of thermal expansion is about half that of its liquid state. Hence, according to IS, the height of the solid mercury column *also* represents temperature and therefore this is not a case of misrepresentation. It just happens that the manufacturer's calibration of the thermometer assumes that normal circumstances prevail.

Let us consider another case. Place a mercury thermometer in a microwave oven and switch it on. The microwave radiation induces an electric current in the mercury column, which heats it. The mercury column expands until it vaporizes at 350 °C. The microwaves do not directly heat the air inside the oven. Hence, in these circumstances, the thermometer fails to measure ambient temperature. Is this a case of misrepresentation?

The thermometer is not being used as intended by its designers. But IS is independent of the intentions of designers. In this setup, the height of the mercury column is reliably caused by the temperature of the mercury or, conjunctively, by the duration the oven has been switched on, or with the integral of the microwave energy supplied, etc. Hence, according to IS, the thermometer's mercury column may correctly represent any of these features.

Fodor (1990, p. 42) identifies the general problem: if representation X refers to Y in virtue of Y reliably causing X then it follows that X "reliably" or "truthfully" represents Y. But a theory of semantic content should be able to explain the possibility of representational error and, therefore, the conditions for X to be a representation of Y must be somehow *separable* from the conditions for X to be a veridical representation of Y. In IS these conditions are identical and therefore IS cannot account for representational error.

Purported examples of misrepresentation turn out to be cases of representation according to IS. So all the different situations in which a Y reliably causes X are a referent of X. IS says that X refers to Y_1 *or* Y_2 *or* Y_3 *or* Y_4, etc.—that is the *disjunction* of all the possible reliable causes of X in different circumstances.

The conjunction and disjunction problems are well-known classic counterarguments to IS that generate semantic indeterminacy. The conjunction problem highlights synchronic indeterminacy (i.e., indeterminacy in a given situation) and the disjunction problem highlights diachronic indeterminacy (i.e., indeterminacy due to multiple situations). In consequence IS admits too many candidates for semantic content and therefore fails to explain how reference is focused. The disjunction problem also entails that IS cannot explain misrepresentation.

Many candidate Ys are admitted by conjunction and disjunction problems. I now give a further, and novel, argument that demonstrates that IS also fails to provide sufficient reasons why a substate X refers to *any* of these Ys. In other words, IS does not even explain how referential content is possible.

15.5 The Problem of Semantic Dualism

Consider again the mercury thermometer. First, let us check it is working. Stick it in a bowl of ice and wait for thermal equilibrium. The thermometer reads $0\,°C$. Now submerge it in a bowl of boiling water. The mercury column rises until it reaches $100\,°C$. All is well.

Now imagine a man from the moon who descends to Earth. He knows a little (moon) science but is entirely ignorant of human practices, language, and notation. He finds an artifact lying on the ground (it is the thermometer we checked but he does not know that). The symbol "°C" written on its face means nothing (it looks like "*♣*" to his eyes). He is curious and therefore starts to experiment on the artifact and, after hours of testing and study, he decides the artifact has something to do with temperature. So he sticks it in a bowl of ice. It reads $0\,°C$. He then submerges it in a bowl of boiling water. It reads $100\,°C$.

He now has the Aha! moment: he reasons that the symbol "°C" is a unit of *length* since it appears beside height marks. He conjectures that the Earth artifact is a device for measuring the coefficient of thermal expansion of mercury, which he calculates is $\frac{100}{x}\,°C/°M$ (that is, Earth units of length per moon units of temperature), where $x\,°M$

is the difference between the boiling and freezing points of water measured in moon units of temperature. (And from an anthropological point-of-view he is delighted because now, via the objective properties of mercury, he can translate between Earth units of length, °C, and moon units of length!)

Where did the man from the moon go wrong? He did not. His conjecture was consistent with the evidence. His only "mistake" was to interpret the thermometer's scale to be a local measure of mercury height rather than an ambient measure of temperature. He didn't know that humans interpret the scale as *referring* to temperature.

The °C scale represents temperature when the artifact is used to measure temperature. But the °C scale represents length when the artifact is used to measure the thermal expansion of mercury. It just so happens that (unbeknown to the man from the moon) humans normally use the artifact as a thermometer and, in human notation, °C in fact means units of temperature.

Hence the relation of reliable causation between the local temperature and the height of the mercury column is *essentially* semantically indeterminate: the very same causal relation, in itself, is consistent with the mercury's height referring to temperature *and* referring to nothing, i.e., merely "representing" itself. The fact that "°C" is conventionally interpreted as a unit of temperature depends on the intentions and practice of the thermometer's users. So the semantic properties of the thermometer are not reducible to reliable causation.

But perhaps this is too hasty? The man from the moon could have considered other factors, such as the thermometer's overall design. The bulb at the end of the capillary tube gives a clue to its true function. The mercury is sealed, which means the device could not be used to measure the thermal expansion of other liquids, which seems an odd limitation. Also, he discovered the artifact in a greenhouse filled with plants.

It's true: all this context, like pieces of a puzzle, helps identify the normal function of the artifact. Yet it would remain a fact that the artifact may be used to measure the thermal expansion of mercury. And when used in this way, the height of the mercury column represents itself—and not something else.

In summary, the thermometer's information-bearing substate of a thermometer has at least two related, but distinct, semantic contents: indirect reference to temperature or direct reference to the height of the mercury column. It just so happens that our standard practice with respect to thermometers hides the direct reference. The identical relation of reliable causation between the thermometer and its circumstances is consistent with either semantic content. The conclusion follows: the semantic content of the thermometer's information-bearing substate, what "it measures", is not determined by IS. Hence there must be some other cause of semantic content not reducible to reliable causation.

Are mercury thermometers special in this respect? Consider a digital thermometer with a built-in platinum thermistor. The thermistor's electrical resistance covaries with temperature. An internal CPU samples the resistance and converts the data to °C for display on a LED screen. Again, assuming ignorance of human ways, does the number on the LED screen represent temperature or electrical resistance? The answer depends on whether the device gets used as a thermometer or as a means for determining platinum's "temperature coefficient of resistance".

Perhaps we can reverse-engineer the software code that converts resistance to '°C' in order to fix the semantics without reference to how the artifact is used? No because the code only reveals that electrical resistance is mapped to an output scale. But what that output scale represents is precisely the question. And if the manufacturer accidentally littered an obscure section of chip memory with some debug symbols, such as

```
float convertToTemperature ( floatresistance )
```

that would help decide normal use. But it would not alter the fact that the artifact can be used to measure platinum's temperature coefficient of resistance.

What's happening here? When we read a thermometer we "look through" the height of the mercury column as if it were a transparent window onto temperature. But we can also "look at" the mercury column as if it were an opaque state caused by temperature. The transparent and the opaque semantic contents are necessarily dual to each other. Let's call this "semantic duality" (SD):

> (SD) Y is a reliable cause of the state of X. Then IS claims that X represents Y (transparent content). But X may also "represent itself" as reliably caused by Y (opaque content) in virtue of the identical causal relation.

How general is SD? Consider a clock. Obviously the transparent content is time. The opaque content, for an analog clock, is the angle of rotation of the clock's hands. A clock is internally driven by its timekeeping element (e.g., pendulum, quartz crystal, etc.) The clock's scale therefore represents time, if we use it to measure time, or angle of rotation, if we use it to measure the degree of hand rotation per tick (e.g., per swing, per oscillation, etc.) of the timekeeping element.

Thermometers and clocks are not that simple. Perhaps very simple measuring instruments lack a dual reading? Consider a ruler. The transparent content is the length of any adjacent object (a ceramic tile, say). The opaque content is the quantity of segments in the ruler's body. So, whether the ruler's scale represents tile length or ruler segments depends on whether we want to measure the length of the tile or the quantity of ruler segments per tile.

Is this last example forced? No. Measuring the "quantity of ruler segments per tile" is the kind of measurement that metrologists perform to calibrate rulers to a standard scale. For example, until recently, the "meter" was defined as the length of a standard metal bar stored at constant temperature. So to calibrate and mark a two meter ruler a metrologist measures the "quantity of ruler segments per metal bar", which, in this case, would be two segments. (Often the opaque semantic content of a measuring device is related to a calibration use-case.)

The examples can be multiplied endlessly (e.g., barometers, accelerometers, compasses, voltmeters, spirit levels, weather vanes, sundials, etc.) IS cannot pick out what these devices actually measure because SD implies that a given relation of reliable causation supports both transparent and opaque content. In fact, whether a thermometer measures temperature (transparent content) or the height of a mercury column (opaque content) depends on the user's intentional choice to either measure temperature or the thermal expansion of mercury. The "crude causal theory", therefore, does not break out of the "intentional circle" and fails to explain why a substate X should possess transparent (i.e., truly referential) content at all.

15.6 Loop-Closing Semantics

Now that we understand some of the problems that any reductive account of semantic properties must face let's turn our attention to Sloman's loop-closing theory.

Tarski (1956) showed that an axiom system S, expressed in a formal language L, admits certain semantic interpretations while excluding others. Tarski's idea was to set up a mapping between formulae in L and structures in a domain of interpretation M. We may then interpret a formula F as "referring" to or denoting a structure in M.

In Tarskian semantics F has the semantic value "true" if it happens to denote a truth in M. The domain of interpretation, M, is therefore called a "model" of an axiom system if each of the axioms denotes a truth in M. For example, the set of natural numbers is a "model" for Peano's axioms of arithmetic.

In general, axioms system admit multiple different interpretations. As Hilbert famously remarked "instead of points, lines, and planes one could consider an interpretation of geometry in terms of 'tables, chairs and beer mugs'" (Feferman and Feferman 2004, p. 279). Hence, what a formula F possibly "represents" depends on our intentional practice of setting up mappings between L and M. Tarski's theory is therefore not a reductive explanation of semantic properties (and was not intended to be). Nonetheless Sloman takes it as a starting point for constructing a reductive explanation of semantic content.

Sloman proposes we generalize Tarski's theory to (i) encompass nonformal "languages" or representational systems, S, implemented as part of the cognitive machinery of an autonomous agent acting in a world; and (ii) consider the causal links between an agent (that uses S to control its actions) and its world. The main idea is that an agent's information-bearing substates S admit "Tarskian" interpretations in terms of structures of an environment E (where S and E are analogous to a logical formula F and a structure in M in Tarskian semantics). But "the existence of causal links removes, or reduces, the ambiguity inherent in the purely [Tarskian] structural semantics … For example, electronic mechanisms ensure that bit patterns in a computer are causally related to locations in its own memory rather than locations in another machine, despite having the same structural relations to both" (Sloman 1997).

Clearly, generalizing a highly developed meta-mathematical theory to include all kinds of representational systems in dynamic interaction with all kinds of environments is no small task. In fact, it defines a whole research program. Sloman (1997) therefore offers us a "thumbnail sketch", which is worth quoting in full since it is the essential programmatic statement of loop-closing semantics:

Consider an environment E containing an agent A, whose functional architecture supports belief-like and desire-like substates. Suppose A uses similar substructures for both, just as a machine can use bit patterns both for addresses and for instructions. Then we can define the class of possible "loop-closing" models for a set of structures S by considering a set of possible environments E satisfying certain conditions, when the action-producing mechanisms, the sensors, and the correspondence tests are working normally:

(a) States in E will tend to select certain instances of S for A's belief-like substates.

(b) If Si is part of a desire-like state of A and E is in state Ei, A's correspondence tests show a discrepancy between Ei and Si, then (unless A's other belief-like and desire-like states interfere) A will tend to produce some environmental state Ej in E which tends to pass A's "correspondence" test for Si.

(c) If that happens Ej will tend to produce a new belief-like state in A.

I repeatedly say "tend to" to indicate that there are many additional factors that can interfere with the tendency, such as conflicts of desires, perceptual defects, accidents, wishful thinking, bad planning, and other common human failings. So these are very loose regularities, and cannot be taken to define internal states in any precise way. None of this presupposes that A is rational. It merely constitutes a partial specification of what belief-like and desire-like mean. However, a full specification will be relative to an architecture, within which functional roles can be defined more precisely.

A number of key ideas are worth noting in this passage.

First, a loop-closing model is an environment in which agent A "tends to" achieve a goal. Internal structures S play two kinds of functional role during this episode. Belief-like states are caused by features in the world. Internal "correspondence tests" compare desire-like states with belief-like states. Discrepancies cause the agent to act on the world until the tests are passed. A mind, on this account, is therefore a generalized feedback control system, a point Sloman repeatedly emphasizes (e.g., (Sloman 2002b)).

Second, in Tarskian semantics the causal links between formula F and a structure in model M are instantiated by our (normally meta-mathematical) practices; but in loop-closing semantics the links between substates S and environment E are instantiated by the agent's practical activity. The agent—as an autonomous, goal-directed mechanism—creates "causal links" between its representations and referents. An independent observer, or "truth-maker", such as ourselves, which sets up and maintains the mapping, is not required.

This thumbnail sketch, of necessity, leaves many questions unanswered. In particular, the sketch does not pick out which *subset* of "states in E" are the semantic content of the agent's internal states. If the "states in E" are a conjunction of features then loop-closing semantics is subject to the conjunction problem. Sloman observes that the causal links are "loose regularities", perhaps due to "perceptual defects", and hence the internal states cannot be defined "in any precise way". This suggests that loop-closing semantics is subject to the disjunction problem, and will therefore fail to identify cases of misrepresentation. We therefore need to add more detail to Sloman's sketch in order to understand how it avoids the kind of semantic indeterminacy that derailed the "crude causal theory" IS.

As we develop Sloman's sketch, I pay repeated attention to a very simple example of a feedback control system, a thermostat coupled to a heating system. This will help make some abstract ideas more concrete.

15.7 Thermostats

The thermostat is a familiar household device that embodies the principles of negative feedback control (e.g., Weiner (1975, pp. 96–97). Users set the thermostat's "set-point" to the desired temperature. A thermometer-like component of the thermostat measures room temperature. The thermostat compares its set-point to the measured temperature. If the set-point is higher the thermostat turns heating on; but if the set-point is lower the thermostat turns heating off (or—if it has the capability—turns cooling on). The thermostat heats or cools the room until the temperature equals the set-point.

To be more concrete, we will keep in mind a specific thermostat design where the thermometer component is a spiral bimetal coil, composed of two metals with differing coefficients of thermal expansion, which winds smaller or unwinds larger as the temperature changes. Moving the set-point rotates the spur of the bimetal coil, which physically tips a bulb containing mercury. Mercury flows to one end of the bulb and completes the heating circuit. As room temperature rises the bimetal coil unwinds. Heating stops when the unwound coil untips the bulb and breaks the circuit.

McCarthy (1979) recounts the story of a heating engineer he called to his home.

> Recently it was too hot upstairs, and the question arose as to whether the upstairs thermostat mistakenly believed it was too cold upstairs or whether the furnace thermostat mistakenly believed the water was too cold. It turned out that neither mistake was made; the downstairs controller tried to turn on the flow of water but could not, because the valve was stuck. (McCarthy 1979)

These intentional ascriptions—"believed", "tried", "mistake"—are useful especially "if we consider the class of possible thermostats, then the ascribed belief structure has greater constancy than the mechanisms for actually measuring and representing the temperature" (McCarthy 1979). Dennett (1997) also emphasizes the utility of viewing the thermostat "as if" it had belief-like and desire-like states that refer to temperature.

In what follows we seek to understand, not the utility, but the actuality of these ascriptions; that is, can loop-closing semantics answer the question of whether a thermostat's information-bearing substates *in fact* refer to temperature, regardless of what we may think, or how we talk about or interact with it. In other words, in contrast to Dennett's approach, we seek to identify how semantic properties necessarily emerge from certain classes of causal structures.

First, a preliminary assumption. I assume that the identification of a thermostat (and, more generally, any control mechanism) is unproblematic. So I rely on some unstated classificatory schemes, mereological principles, and concepts of boundary—specifically distinctions between control mechanisms and their parts and the worlds they inhabit. For example, our definition of a thermostat includes the heating component, such as a home furnace, as part of the thermostat, but not the air molecules close to the heating element or the home in which it functions.

I also assume that class instances, such as the specific thermostat in your home, remain exemplars of their class in all the varied circumstances we consider. I exclude broken or malfunctioning thermostats and cases in which some other agency in the world (such as child with a screwdriver) directly interferes with the inner workings. Only when we have a theory of the semantic properties of a class of mechanisms, modulo the worlds in which they function, can we begin to think of more general cases in which the actual mechanism itself is subject to change. So in what follows I fix the definition of control mechanism, keeping it constant, while varying its environment.

15.8 The Metalanguage of Causal Graphs

To expand Sloman's sketch, we need clarity on kinds of "causal links". I use causal graphs for this purpose, borrowing from manipulability theories of causation, especially as developed by Pearl (2000) and Woodward (2003). Adopting the formalism of casual graphs is merely a pragmatic choice for this chapter. The more important point is that, just as Tarskian semantics requires a "metalanguage" to describe both the formal system and its model, we also need a metalanguage to describe control mechanisms, their worlds and the causal relations between them.

A causal graph captures the fixed causal structure of some part of a world at a given level of description. It is a directed graph represented by the ordered pair (V, E), where V is a set of vertices that denote variable properties of ontological features of arbitrary type (e.g., ambient temperature, height of mercury column, state of a data structure, etc.) and E is a set of pairs denoting directed edges between vertices that represent invariant causal relationships between those features (e.g., temperature is a direct cause of the height of the mercury column, etc.) (Pearl 2000). If we are uncertain about the causal relationship between two features we connect them with an undirected edge that represent covariation (e.g., is a metal's temperature the cause of its thermal expansion or identical with it?) Each variable $v_i \in V$ is some arbitrary function of its direct parents. So we associate a set of functions, $F = \{v_1 = f_1(\cdot), v_2 = f_2(\cdot), \ldots\}$, with every causal graph.

15.9 Control Mechanisms

We define the abstract causal structure of a class of control mechanisms in terms the following causal graph.

Fig. 15.1 A causal graph of a
control mechanism $\mathscr{K}$

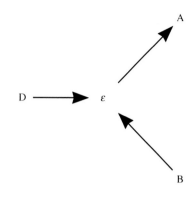

Definition 1 A *control mechanism* $\mathscr{K}$ is the causal graph

$$(V_{\mathscr{K}}, E_{\mathscr{K}}) = (\{B, A, \varepsilon, D\},$$
$$\{(B \rightarrow \varepsilon), (D \rightarrow \varepsilon), (\varepsilon \rightarrow A)\})$$

where B is a belief-like substate, A is an action-like substate, ε is an error-like substate, and D is a desire-like substate (see Fig. 15.1). The associated function set satisfies

$$\varepsilon = 0 \iff A = \perp,$$

where $\varepsilon = 0$ means that $\mathscr{K}$'s 'correspondence test' between D and B is satisfied and $A = \perp$ means the 'null action' (i.e., control mechanism $\mathscr{K}$ is not performing an action and therefore A is not the direct cause of any event).

My use of the terms belief-like, desire-like, etc. do not imply the information-bearing substates have semantic content (since that has yet to be established). The terms are simply shorthand for their functional role within the control mechanism.

For example, interpret Fig. 15.1 as depicting the causal relations in a thermostat. Then variable D represents the thermostat's set-point (say $D = 72\ °C$); variable B is a winding measure of the thermostat's spiral bimetal element; variable A is a measure of the furnace's heat output in Watts; and variable ε, for 'error', measures the discrepancy between B and D, which is the distance between the spur of the spiral coil and the mercury bulb.

Graph $\mathscr{K}$ simply states that B and D are direct causes of ε, and ε is the direct cause of A. Our metalanguage description of control mechanism $\mathscr{K}$ can therefore represent other control mechanisms that share the same casual structure, such as servomechanisms, or (perhaps transient) control mechanisms in more sophisticated information-processing architectures that execute in higher level virtual machines.

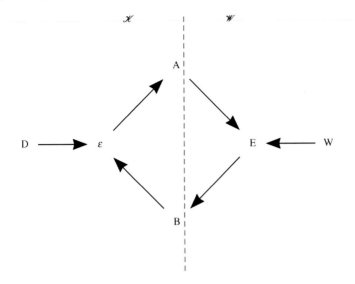

Fig. 15.2 A loop L formed by embedding control mechanism $\mathcal{K}$ in a world $\mathcal{W} = (\{E, W\}, \{(W \to E)\})$. The *dotted line* represents the boundary between the control mechanism and its world

15.10 Worlds, Loops and Models

A world is any arbitrarily complex causal graph that doesn't include the individual control mechanism we're considering.

Definition 2 A *world* $\mathcal{W} = (V_{\mathcal{W}}, E_{\mathcal{W}})$ is any causal graph such that $V_{\mathcal{W}} \cap V_{\mathcal{K}} = \{\}$.

A control mechanism $\mathcal{K}$ is open to inputs (which affect its belief-like state B) and can emit outputs (which are caused by its action-like state A). We can therefore form a control loop by connecting $\mathcal{K}$ to a world $\mathcal{W}$ in the 'right way'.

Definition 3 A *loop* $L_{\mathcal{K},\mathcal{W}} = (V_L, E_L)$ is the directed cyclic graph formed by embedding control mechanism $\mathcal{K}$ in world $\mathcal{W}$ such that

(a) $V_L = V_{\mathcal{K}} \cup V_{\mathcal{W}}$,
(b) $E_L = E_{\mathcal{K}} \cup E_{\mathcal{W}} \cup E_*$, where $(A \to v_i) \in E_*$ and $(v_j \to B) \in E_*$ for some $v_i, v_j \in V_{\mathcal{W}}$,
(c) and there is at least one path from A to B.

Assume that all loops are empirically realizable.

Figure 15.2 depicts a loop formed by embedding $\mathcal{K}$ into a very simple world $\mathcal{W}$ consisting of two variables. Interpret variable E, the direct 'environment', to indicate a feature of the world directly connected to $\mathcal{K}$ and interpret variable W, 'wild causes', to indicate the effect of some other causal agent in the environment. (The functional relations between the variables are not shown, e.g., $E = f_1(A, W)$, $B = f_2(E)$ etc.)

A loop has implicit feedback dynamics in virtue of the functional relations between its variables. Since we make no restriction on the kind of functional relations (they could be arbitrary programs) loops can therefore instantiate arbitrary dynamic systems (whether continuous or discrete, deterministic or probabilistic, symbolic etc.) The only restriction is a fixed causal structure (although finite-sized variable causal structures can be represented by functions that switch subgraphs in and out).

A model for a control mechanism $\mathcal{K}$ is any world in which it can form a loop.

Definition 4   A *model* is any world $\mathcal{W}$ that can form a loop $L_{\mathcal{K},\mathcal{W}}$ with control mechanism $\mathcal{K}$.

15.11 Inactivity

For example, Fig. 15.3 depicts a thermostat in a room with an open window. Call it model $\mathcal{A}$. The functional relations in this graph are fully specified as differential equations.

In Tarskian semantics the term 'model' is reserved for those domains of interpretation that satisfy a given logic. In loop-closing semantics we reserve the term 'loop-closing model' to denote the subset of models in which control mechanism $\mathcal{K}$ is (i) ultimately inactive and (ii) causally responsible for its inactivity. Let's examine what these conditions mean.

A trajectory in loop $\mathcal{L}_{\mathcal{K},\mathcal{W}}$ is any time-ordered trajectory in state-space generated by the loop's dynamics.

For example, the function set of model $\mathcal{A}$ is $F = \{\varepsilon = D - B, A = \alpha\varepsilon, \frac{dE}{dt} = A + \gamma W, B = \beta E\}$ (see Fig. 15.3). In this specification D, W, α, γ and β are free variables. So model $\mathcal{A}$ in fact defines a family of models. Let's examine two fully specified models from this family.

Loop $\mathcal{A}_1$, with $D = 72$, $W = 0$, $\alpha = 1$, $\gamma = 50$, and $\beta = 1$, represents a thermostat with a fixed set-point in a room with a permanently closed window.

Loop $\mathcal{A}_2$, with $D = 72$, $W \approx |\sin t|$, $\alpha = 1$, $\gamma = 50$, and $\beta = 1$, represents a thermostat with a fixed set-point in a room with a window that continually opens and closes (modeled by the $\sin t$ term).

Figure 15.4 plots both loops given some initial conditions. We immediately see that the thermostat in loop $\mathcal{A}_1$ 'successfully' controls the room temperature whereas the thermostat in loop $\mathcal{A}_2$ fails due to the variable cooling caused by the open window (the thermostat is always playing 'catch up'). The thermostat is 'successful' in the sense that it enters a state of inactivity, or quiescence, in which its action-like state A is not the direct cause of any event.

Definition 5   A control mechanism $\mathcal{K}$ is *inactive* when $\varepsilon = 0$; otherwise, it is *active*.

For example, an active thermostat heats the room (i.e., $A \neq \bot$), which causes its temperature-sensing element to change state. But when the room temperature equals

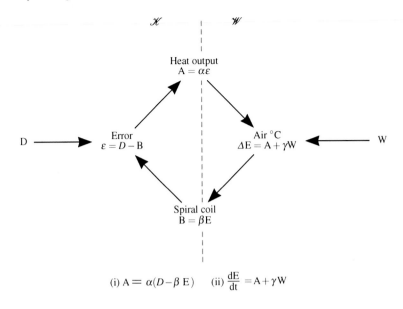

$$\text{(i) A} = \alpha(D - \beta\,E) \qquad \text{(ii) } \frac{dE}{dt} = A + \gamma W$$

Fig. 15.3 Loop $\mathscr{A}$: a loop formed by a thermostat embedded in a room with a window. D is the thermostat's set-point and W is the area of the window open to the outside air. The functional relations between the variables define a simple dynamic feedback system (Eqs. (i) and (ii))

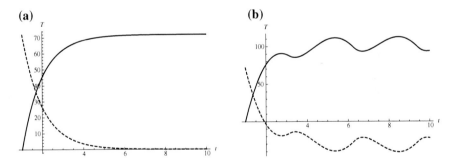

Fig. 15.4 State-space trajectories of thermostat in a room with a window. The *solid line* represents room temperature and the *dotted line* represents the thermostat's heat output

the set-point then heating ceases and (for some nonzero period) the thermostat neither heats or cools (i.e., $A = \perp$).

Note that $\varepsilon = 0$ implies $A = \perp$ by the definition of a control mechanism. Inactive control mechanisms are 'in equilibrium' with their environment because the environment corresponds to their desire-like substate.

Definition 6 A trajectory is *ultimately inactive* if for some t_0 (i) control mechanism $\mathscr{H}$ is inactive and (ii) for all $t < t_0$ control mechanism $\mathscr{H}$ is active.

The first condition in this definition requires that the control mechanism enter a state of equilibrium or inactivity. The second condition simply excludes degenerate trajectories where $\mathscr{K}$ is never active.

In our simple examples, we could use standard Lyapunov stability theory to prove that model $\mathscr{A}_1$ is ultimately inactive (for all d_0) while model $\mathscr{A}_2$ is not. In general, however, such proofs elude us and we must rely on finite, and therefore fallible, observation of the trajectory itself.

A trajectory may be ultimately inactive yet $\mathscr{K}$ need not be the causal agent solely responsible for it. The outcome could be fortuitous, accidental, or necessarily depend on the actions of some other agent. For example, consider a variant of model $\mathscr{A}_2$ in which a standalone heating lamp synchronizes its output with the opening and closing of the window. The heat lamp negates the cooling effect of the open window and the loop is then ultimately inactive. But here $\mathscr{K}$ is not 'sole cause' of its own inactivity. Let's turn, therefore, to identifying the conditions in which $\mathscr{K}$ is causally responsible.

15.12 Output as the Actual Cause of Input

An active control loop, in which $\mathscr{K}$ repeatedly emits actions and senses the consequences, may be initiated by a change in its desire-like or its belief-like substates.

In some loops $\mathscr{K}$ is 'in control' because its action A fully determines those features of its environment measured by B. For example, if $\mathscr{K}$ is a thermostat in a closed room that contains no other sources of heat then, in all likelihood, it will fully control the ambient temperature and hence the winding of its spiral coil. In contrast, if cold air blows through an open window, or a heating engineer directs a hair-dryer toward the thermostat to check it's working, then the thermostat is not fully 'in control'. In these cases the thermostat either partially controls or fails to control its own belief-like substate B. Can we make these kinds of distinctions precise?

The concept of 'actual cause' employed in manipulability theories of causation (Woodward 2003) specifies what it means for $\mathscr{K}$ to be 'in control'. The actual cause is an event 'recognized as responsible for the production of a given outcome in a specific scenario, as in "Socrates drinking hemlock was the actual cause of Socrates death"' (Pearl 2000, p. 309). Socrates executed his own death sentence by voluntarily drinking hemlock because he believed the law should be obeyed. His belief was a 'contributing cause' of his death. But it was the ingestion of hemlock that actually killed him.

Pearl's elegant formal theory of causal intervention provides precise definitions of 'actual cause' and related notions (Pearl 2000). But to maintain a focused exposition I will use a relatively informal definition (borrowing from the formal theory) sufficient for our purposes.

First, we need the idea of an intervention. An 'intervention' is intended to capture 'processes that satisfy whatever conditions must be met in an ideal experiment designed to determine whether X causes Y' (Woodward 2003, p. 46).

Definition 7 I is an *intervention variable* for X (the 'causing variable') with respect to Y (the 'effect variable') if and only if (i) I is a direct cause of X, (ii) I is the only direct cause of X, (iii) any directed path from I to Y goes through X, and (iv) if there is a directed path from any other variable V to I that does not go through Y, then any directed path from V to Y goes through X (this definition based on Woodward (Woodward (2003), pp. 98–100)).

Consider the causal graph depicted in Fig. 15.2. To illustrate the concept let's determine whether ε is an intervention variable for action A with respect to belief B. We can easily see that (i) ε is the direct cause of A, (ii) ε is the only direct cause of A, and (iii) any path from ε to B goes through A. Hence, the first three conditions are met. Condition (iv) is more complex. It's designed to exclude cases where a variable V affects both the 'causing variable' X and the 'effect variable' Y but does so via an independent route. If such a V existed then an intervention on X could not be the unique cause of an effect on Y. The causal structure of a control mechanism $\mathcal{K}$ in fact guarantees there cannot be any such V. So condition (iv) is also satisfied and hence ε is an intervention variable for A with respect to B.

Intervention variables support intervention events, or simply 'interventions'.

Definition 8 I taking some value $I = z_i$ is an *intervention* on X with respect to Y if and only if I is an intervention variable for X with respect to Y (Woodward 2003, p. 98).

In other words, if I is an intervention variable then I taking a new value z_i is a special kind of event called an intervention. We can now define our target concept, 'actual cause'.

Definition 9 $X = x$ is an *actual cause* of $Y = y$ if and only if there is at least one path R from X to Y for which an intervention on X will change the value of Y given that other causes Z_i of Y, which are not on path R, are fixed by interventions at their actual values (cf. Woodward (2003, p. 77)).

For $X = x$ to be the actual cause of $Y = y$ then X and Y must be causally connected and other possible extraneous or 'off path' causes of Y are either absent or fixed at constant values. When X is the actual cause of Y then changes in the value of Y are ultimately and uniquely traceable to changes in the value of X.

Note that the definition of 'actual cause' is counterfactual: it does not require that an exogenous intervention on X actually take place. Hence, X taking the value x can be the actual cause of Y taking the value y in the context of a self-sustaining loop of causation in which X is itself endogenously caused (such as a loop in which $\mathcal{K}$ is active).

We can now define what it means for a control mechanism to be 'in control'.

Definition 10 A control mechanism $\mathcal{K}$ is *loop-controlling* when its action-like substate $A = a$ is the actual cause of its belief-like substate $B = b$ (via some 'active path' R from A to B).

A control mechanism is loop-controlling, therefore, when its output is the actual cause of its input.

Let's return to the variant of model $\mathscr{A}_2$ where a heating lamp contributed to the thermostat's ultimate inactivity. In this case the thermostat is not loop-controlling because the lamp is another cause of the winding or unwinding of its spiral coil. So although the loop is ultimately inactive the control mechanism is not the actual cause of this outcome.

A control mechanism may fail to be loop-controlling even in the absence of 'wild factors' such as W. For example, consider a thermostat that heats water in a container. Set the thermostat's set-point to $D = 105$ °C. Water boils at 100 °C at which point its temperature stabilizes during vaporization. At this temperature the thermostat's output is also not the actual cause of its input (since any increase in heating output does not change the winding of its spiral coil).

The definition of loop-controlling handles not just simple cases but scales to arbitrarily complex causal graphs.

Now that we've defined situations when a mechanism controls its own input we can consider complete dynamic trajectories where the control mechanism is fully 'in control'.

15.13 Loop-Closing Models

To 'close' a loop means to terminate its activity. A control mechanism implicitly defines a set of 'loop-closing models'. A loop-closing model is a world in which it's possible for the control mechanism to be the actual cause of its own inactivity.

Definition 11 A *loop-closing trajectory* is (i) an ultimately inactive trajectory in which (ii) control mechanism $\mathscr{K}$ is loop-controlling when active.

Definition 12 A world $\mathscr{W}$ is a *loop-closing model* for control mechanism $\mathscr{K}$ if there exists at least one loop-closing trajectory in loop $\mathscr{L}_{\mathscr{K},\mathscr{W}}$.

A loop-closing model is like a 'natural' controlled experiment that creates conditions in which $\mathscr{K}$ manifests its casual powers without interference. Controlled experiments are designed 'to get a single mechanism going in isolation and record its effects' since outside of experimental conditions the powers of the mechanism 'will normally be affected by the operations of other mechanisms' such that 'no unique relationship between the [observed] variables or precise description of the mode of operation of the mechanism will be possible' (Bhaskar 1997, p. 43). Any countervailing mechanisms that could affect the causal links between $\mathscr{K}$'s outputs and inputs are either absent or inactive—and therefore 'controlled for'. The loop-closing models are experiments that an observer, ignorant of the function and design of the control mechanism, might need to setup in order to identify its essential causal powers.

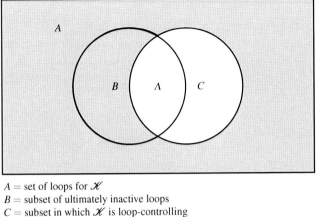

A = set of loops for $\mathcal{K}$
B = subset of ultimately inactive loops
C = subset in which $\mathcal{K}$ is loop-controlling
$\Lambda = B \cap C$ = subset in which $\mathcal{K}$ is the actual cause of its own inactivity
$\Lambda_{\mathcal{K}} = \{ \mathcal{W} \mid L_{\mathcal{K},\mathcal{W}} \in \Lambda \}$ = the set of loop-closing models.

Fig. 15.5 Venn diagram representation of the set of loop-closing models

An analogy might help here. Imagine you find a door key. This key opens some subset of all the locks in the world. The properties of the key —its size, shape, and number of teeth, etc,—implicitly define the set of locks it can open. In principle, you can enumerate this set by successfully unlocking doors (on condition you control for countervailing and interfering factors, such as rusty locks, broken mechanisms, hidden secondary locks, etc.) The loop-closing models, then, are worlds that control mechanism $\mathcal{K}$ 'unlocks' in virtue of its causal powers alone and therefore $\mathcal{K}$, in these circumstances, is the actual cause of its own inactivity. Figure 15.5 depicts the logical relationship between the set of all possible loops and the special subset of loop-closing models.

Loop-closing models 'define a non-Tarskian model for the internal representations which play a role in percepts, beliefs plans, etc., namely an external environment, which can coherently close the feedback loops' (Sloman 1986b). We've deliberately restricted our discussion of control mechanisms to the relatively simple causal structure $\mathcal{K}$. But in general 'this notion of coherent causal closure will be relative to the system's ability to have precise and detailed goals and beliefs. How specific the mapping is between internal representations and external structures will depend on how rich and varied is the range of percepts, goals, and action strategies the system can cope with' (Sloman 1986b).

Let's turn now to constructing a Tarksian-like mapping between internal representations and the external structures of loop-closing models.

15.14 The Manipulable Feature

A control mechanism $\mathcal{K}$ controls some special feature or collection of features in virtue of its causal powers. I call this collection of features the 'manipulable feature'. The set of loop-closing models possess the remarkable property of implicitly specifying what the manipulable feature actually is. Let's now make that explicit.

Definition 13 The *conjunctive feature* of a loop-closing model, $\mathscr{W} = (V_{\mathscr{W}}, E_{\mathscr{W}})$, is the conjunction

$$f_{\mathscr{W}} = f_1 \wedge f_2 \wedge \cdots \wedge f_n,$$

where (i) $f_i \in V_{\mathscr{W}}$ and (ii) there exists at least one loop-closing trajectory in $\mathscr{L}_{\mathscr{K},\mathscr{W}}$ with f_i on its active path.

The conjunctive feature is simply the conjunction of features that connect $\mathcal{K}$'s outputs to its inputs in a loop-closing trajectory.

For example, imagine we attach a thermostat to the Rube Goldberg setup we described in Sect. 15.3. Figure 15.6a depicts the loop's casual graph, $\mathscr{L}_{\mathscr{K},\mathscr{W}_1}$. Assume the functional relations entail that $\mathscr{W}_1$ is loop-closing. The conjunctive feature is then

$$f_{\mathscr{W}_1} = \text{Liquid } ^{\circ}\text{C} \wedge \text{Color} \wedge \text{Engine} \wedge \text{Aperture} \wedge \text{Air } ^{\circ}\text{C},$$

which is simply the conjunction of features on the active path, where these labels denote variable properties (e.g., 'Engine' is shorthand for the rotary speed of a thermokinetic engine).

For example, consider a thermostat that controls the temperature of a closed room in which temperature and pressure covary (loop $\mathscr{L}_{\mathscr{K},\mathscr{W}_2}$ in Fig. 15.6b). The conjunctive feature is

$$f_{\mathscr{W}_2} = \text{Air } ^{\circ}\text{C} \wedge \text{Air pressure}.$$

Another example: consider a thermostat that controls the air temperature inside a balloon. The balloon expands and contracts and hence temperature and pressure do not covary (loop $\mathscr{L}_{\mathscr{K},\mathscr{W}_3}$ in Fig. 15.6c). The conjunctive feature in this case is

$$f_{\mathscr{W}_3} = \text{Air } ^{\circ}\text{C}.$$

Finally, consider a thermostat that controls the temperature of a metal wire soldered to the heating element and the spiral coil (loop $\mathscr{L}_{\mathscr{K},\mathscr{W}_4}$ in Fig. 15.6d). The conjunctive feature is

$$f_{\mathscr{W}_4} = \text{Metal } ^{\circ}\text{C} \wedge \text{Metal volume}.$$

The sample of loop-closing models for thermostat $\mathcal{K}$ are then $\Lambda_{\mathscr{K}} = \{\mathscr{W}_1, \mathscr{W}_2, \mathscr{W}_3, \mathscr{W}_4\}$. The models have varied causal structures but share the common property of forming coherent 'causal closures' with $\mathcal{K}$.

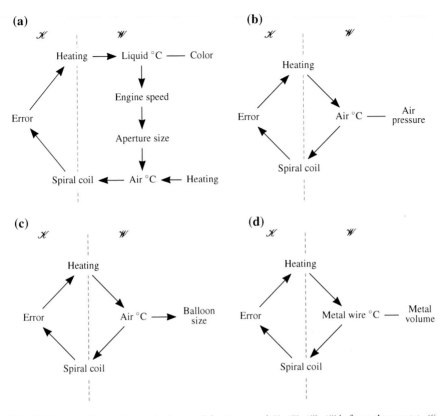

Fig. 15.6 A small set of loop-closing models, $\Lambda_{\mathscr{K}} = \{\mathscr{W}_1, \mathscr{W}_2, \mathscr{W}_3, \mathscr{W}_4\}$, for a thermostat $\mathscr{K}$. **a** $\mathscr{L}_{\mathscr{K}, \mathscr{W}_1}$: controlling the conjunction of features in a Rube Goldberg loop. **b** $\mathscr{L}_{\mathscr{K}, \mathscr{W}_2}$: controlling the air temperature in a room. **c** $\mathscr{L}_{\mathscr{K}, \mathscr{W}_3}$: controlling the air temperature inside a balloon. **d** $\mathscr{L}_{\mathscr{K}, \mathscr{W}_4}$: controlling the temperature of a metal wire

Definition 14 The *disjunctive feature*, $f_{\Lambda_{\mathscr{K}}}$, of a set of loop-closing models, $\Lambda_{\mathscr{K}} = \{\mathscr{W}_1, \mathscr{W}_2, \dots\}$, for control mechanism $\mathscr{K}$, is the disjunction

$$f_{\Lambda_{\mathscr{K}}} = f_{\mathscr{W}_1} \vee f_{\mathscr{W}_1} \vee \dots,$$

where each $f_{\mathscr{W}_i}$ denotes the conjunctive feature of model $\mathscr{W}_i$.

In our example, $f_{\Lambda_{\mathscr{K}}}$ is the expression

$$(\text{Liquid } °C \wedge \text{Color} \wedge \text{Engine} \wedge \text{Aperture} \wedge \text{Air } °C)$$
$$\vee (\text{Air } °C \wedge \text{Air pressure})$$
$$\vee (\text{Air } °C)$$
$$\vee (\text{Metal } °C \wedge \text{Metal volume}),$$

which is simply the disjunction of all the conjunctive features.

For large sets of loop-closing models the disjunctive feature is highly complex and will typically exhibit redundancies. We want to obtain the minimal set of features necessary for loop-closure. (For example, the presence of a thermokinetic engine is neither sufficient or necessary for the thermostat's loop to close.) We now state the key organizing concept of this chapter.

Definition 15 The *manipulable (or controlled) feature* of a set of loop-closing models $\Lambda_{\mathcal{X}}$ is the minimal disjunctive normal form (MDNF) of the disjunctive feature $f_{\Lambda_{\mathcal{X}}}$.[1]

The manipulable feature is the minimal specification of features that, if present on an active path, imply loop-closure is possible. In worlds that lack the manipulable feature the control mechanism cannot be the actual cause of its own inactivity.

We can use the Quine-McCluskey algorithm (McCulskey 1959) to compute the MDNF expression. For example, the disjunctive feature in our example reduces to the manipulable feature

$$\text{Air } °C \vee (\text{Metal } °C \wedge \text{Metal volume}).$$

So we can conclude, given this small sample of loop-closing models, that the thermostat's manipulable feature is air temperature *or* the temperature of a metal wire (and its covarying volume). The satisfaction of at least one conjunctive clause of the manipulable feature is sufficient for loop-closure. Simply put, the presence of air or a metal wire between the thermostat's inputs and outputs is sufficient. The presence of the other Rube Goldberg transmission mechanisms are not required for loop-closure and therefore excluded from the definition of the manipulable feature.

We cannot conclude that the absence of air or a metal wire implies loop-closure is impossible because we haven't considered the set of all possible loop-closing models. The thermostat might be able to achieve loop-closure in worlds we have yet to consider.

Note that, in our example, air pressure is not a manipulable feature because model $\mathcal{W}_3$—controlling the air temperature inside a balloon—excludes it. In general, we need highly specific worlds (i.e., natural or artificial experimental situations) to 'control for' interfering variables and thereby exclude features.

Additional loop-closing models can further constrain the manipulable feature. Consider a variant of loop-closing model $\mathcal{W}_4$ where we clad the metal wire in ceramic housing to prevent its thermal expansion. In consequence the volume of the metal wire is now constant and does not vary with its temperature. The conjunctive feature of this new loop-closing model is

$$f_{\mathcal{W}_4'} = \text{Metal } °C.$$

---

[1] I ignore some technical details here. First, some conjunctive features are necessarily coupled and order-dependent due to the associated functional relations. We can take account of order-dependence but doing so would complicate the exposition without altering the basic argument. Second, the MDNF may not be unique, in which case we have a set of manipulable features.

Add this feature as a new term to the disjunctive feature and apply the Quine-McCluskey algorithm again. The manipulable feature for the expanded set of loop-closing models is

$$\text{Air } {}^{\circ}\text{C} \vee \text{Metal } {}^{\circ}\text{C}.$$

The presence of thermal expansion is therefore unnecessary for loop-closure.

The specification of the manipulable feature is of course relative to the choice of a metalanguage. Although we share the same world as a given control mechanism Sloman remarks 'it would be incoherent to try to describe the common underlying reality in neutral terms' (Sloman 1986a). In consequence, 'Like Tarskian semantics, "loop-closing semantics" leaves meanings indeterminate. For any level of specification at which a loop-closing model can be found, there will be many consistent extensions to lower levels of causal structure (in the way that modern physics extends the environment known to our ancestors), which remain adequate models in this sense' (Sloman 1986b).

For example, the loops depicted in Fig. 15.6 happen to be specified at a 'common sense' level of abstraction. The manipulable feature is 'air temperature' or 'metal temperature'. We can extend the metalanguage to include the theory of heat diffusion and re-describe the loops in $\Lambda_{\mathcal{H}}$. The extended metalanguage identifies the common ontological features of the disjunction (i.e., 'jiggling atoms') and collapses the manipulable feature to the natural kind term 'temperature', suitably restricted to the range of values and kinds of functional relations the thermostat can control. (In general, unrestricted 'temperature' features are not manipulable; for example, a domestic thermostat cannot control the temperature of the sun.)

My guess is that, as we expand the set of models to include all possible worlds, then the manipulable feature will converge to the fixed point 'temperature'. This would be the thermometer's *ultimate manipulable feature*. After all, a thermometer has very specific and dedicated input and output channels specially designed to control temperature.

So by a roundabout route we have arrived at the common sense conclusion that a thermostat in fact controls temperature—but not in virtue of the intentions of its designer, or the uses to which it is normally put, or the reasons for its existence and persistence as an artifact, or the method by which it acquired its causal properties—but *in virtue of the kind of thing it is*. The manipulable feature is an objective property of a control mechanism and its possible control loops.

In general, manipulable features exist without necessarily being fully known, either by an observer, any designer, or user of the mechanism, or the control mechanism itself (even if the mechanism is a component of an intelligent agent, such as ourselves).

15.15 Some Reductive Definitions

The causal structure of control mechanism $\mathscr{K}$ defines the functional roles of its interrelated substates. We imply those functional roles by calling D 'desire-like', B 'belief-like', A as 'intentional action-like', and ε as an 'error', 'comparison' or 'discrepancy' state. Let's now address how loop-closing theory claims that these substates indeed have the semantic properties implied by their functional role names.

Recall that Sloman (1997) states that loop-closing theory 'merely constitutes a partial specification of what "belief-like" and "desire-like" *mean*' (my emphasis). We now have the conceptual tools to *define* semantic properties, such as what a belief-like state 'represents', in terms of nonsemantic causal relations (from the perspective of an observer using a given metalanguage). So now I reintroduce semantic terms in order to reductively define them.

First, we need one further distinction. Consider again loop $\mathscr{A}_2$ where a thermostat heats a room with a window that continually opens and shuts. $\mathscr{A}_2$ is not loop-closing. However, it would be loop-closing if we intervened and fixed the window in position.

Definition 16 A nonloop-closing model $\mathscr{W}$ is a *counterfactual loop-closing model* if it can be transformed into a loop-closing model by a set of interventions that fixes some subset of variables $v \subseteq V_{\mathscr{W}}$ at their initial values.

So $\mathscr{A}_2$, while not loop-closing, is counterfactual loop-closing. In contrast, consider a new Rube Goldberg loop $\mathscr{A}_3$, where we place an inverting contraption between the thermostat's output and input such that increased (resp. decreased) heat output gets inverted to increased (resp. decreased) cooling input. This loop is not loop-closing. Also, the inverting functional relations on the active path cannot be changed by intervention (since the inverting contraption lies on the only active path and we restrict interventions to those that 'hold things constant' rather than modifications to the causal structure). So $\mathscr{A}_3$ is not counterfactual loop-closing.

In counterfactual loop-closing models the manipulable feature is present—it just so happens that other mechanisms prevent the loop from closing. In models that are neither loop-closing or counterfactual loop-closing the manipulable feature is absent.

Now to the reductive definitions. The vertices (or substates) of $\mathscr{K}$ take a range of values of a given type. For example, in the case of a thermostat, the belief-like state B is a scalar winding measure of the spiral coil. We map, in a Tarksian-inspired manner, the state of the information-bearing substates (e.g., the coil is 80 % wound) to the *covarying* state of the manipulable feature (e.g., the temperature is 20 °C).

Definition 17 The *semantic content* of belief-like state B is the covarying state of the manipulable feature (B 'refers' to the manipulable feature). B has the *semantic value* 'true' in all loop-closing or counterfactual loop-closing models; and the value 'false' otherwise.

B is 'true' when the manipulable feature is present and 'false' otherwise. Note that we make no mention of whether control is successful or not. Neither do we mention 'ideal' or 'normal' conditions, or higher order teleological concepts of the proper function of a belief-like state, such as the intended use of an artifact.

Definition 18 The *semantic content* of action-like state A is a 'command' to change the covarying state of the manipulable feature (A is 'directed at' the manipulable feature). A has the *semantic value* 'effective' in loop-closing models; and the value 'ineffective' otherwise.

Definition 19 The *semantic content* of desire-like state D is the absence of the manipulable feature in a specific state (D 'refers' to an unrealized state-of-affairs). D has the *semantic value* 'unsatisfied' when control mechanism $\mathscr{K}$ is active; and the value 'satisfied' otherwise.

And we shall also say that $\mathscr{K}$'s activity is a (simple) kind of 'intentional practice', i.e., goal-directed activity, that is *successful* in loop-closing models, and *unsuccessful* otherwise.

Semantic content is invariant across models since the manipulable feature is a property of the control mechanism. Semantic values, however, vary from model to model, since they are a property of the control mechanism's substates in particular worlds.

To illustrate these definitions let's consider the thermostat again. Loop-closing theory therefore claims that the winding factor of the thermostat's spiral coil represents temperature, its set-point represents a possible temperature, and its heating output is a command to alter the temperature. In loop-closing models, such as example $\mathscr{W}_2$, its belief-like state is true (i.e., the spiral coil veridically represents temperature), its action-like state is effective (i.e., contributes to ultimate success), and its desire-like state is ultimately satisfied.

In loop $\mathscr{A}_2$, in which the window opens and shuts, the thermostat's belief-like state is also true (the spiral coil reliably covaries with the manipulable feature), but its actions are ineffective (i.e., they do not contribute to ultimate success) and its desire-like state is forever unsatisfied.

Let's consider a trickier example. Consider a situation where a thermostat outputs heat into room 1 but measures temperature in room 2, where room 1 and room 2 are thermally isolated from each other. Loop-closing theory states that B is false and A is ineffective even though B is reliably caused by the temperature in room 2 and the temperature in room 1 is reliably caused by A. Does this make sense?

Yes. The semantic content of the thermostat's substates is the *manipulable* feature 'temperature'. It's true that rooms 1 and 2 both have temperatures. But the thermostat cannot manipulate (that is, control) the temperature in room 1 nor the temperature in room 2. The manipulable feature of the world necessarily present for control success is absent. Hence, it's belief-like state is false and its action-like state ineffective.

In this situation we, as observers, can use the winding of the spiral coil to veridically measure the temperature in room 2. But the substate would possess this content in virtue of our intentional practice. In contrast, when the substate is recruited by the thermostat itself, as part of its own 'intentional' activity, then the spiral coil falsely represents a manipulable temperature with a specific state. In this situation, the thermostat's belief-like substate misrepresents the world.

Of course, the thermostat has no concept of temperature. So its semantic content is of the 'nonconceptual' or procedural, rather than declarative, kind. It possesses nothing remotely close to our current theory of heat. Hence it doesn't refer to temperature in the way we do; nonetheless, loop-closing theory claims that thermostats do refer to that same aspect of a 'common underlying reality' that we call 'temperature'.

Definitions 15.17–15.19 reduce semantic to causal properties. Do they constitute a successful reduction? We need to test the theory to answer this question.

15.16 Conjunction, Disjunction, and Semantic Duality

At the start of this chapter, we identified three features of semantic properties—reference, focus, and semantic value—that we wanted to reductively explain, at least for simple mechanisms with internal structures that we fully understand. Crude causal theory IS attempts to reduce semantic properties to reliable causation between a referent Y and reference X. The 'atom of meaning', in this theory, is lawful covariation, such as the lawful covariation between a thermometer's mercury column and temperature.

The 'crude causal theory' encounters various logical problems: The conjunction problem implies that any 'upstream' feature Y can be the referent of any 'downstream' feature X in a chain of reliable causation. So IS cannot explain semantic focus. The disjunction problem implies that IS cannot separate the conditions necessary for representation from those necessary for veridical representation. So IS cannot explain semantic values, such as the truth and falsity of beliefs or the possibility of misrepresentation. And the problem of semantic duality implies that IS is indeterminate with respect to transparent and opaque content. So IS cannot explain reference, that is how a substate represents some thing other than itself.

It's clear that semantic properties cannot be reduced to reliable causation. How does loop-closing theory fare on these issues?

15.16.1 Semantic Focus

The 'atom of meaning' in loop-closing theory is a negative feedback control mechanism. The causal structure of control loops may involve reliable causation (such as heat causing thermal expansion) but 'one-way' cause-and-effect relationships, in themselves, under-determine semantic content. Loop-closing theory avoids the conjunction problem because it identifies semantic content with a specific kind of thing—the manipulable feature—which is necessarily present for control success, but in general may be absent. The manipulable feature may appear anywhere, whether proximate or distal, in a complex Rube Goldberg chain of causation or—even more strikingly—not appear at all.

Semantic focus is ultimately explicable in terms of the limited capabilities of specific feedback control mechanisms with specific internal components and dedicated input and output channels. For instance, a thermostat's manipulable feature is temperature (and not altitude), a cruise control system's manipulable feature is velocity (and not temperature), and an automatic inventory system's manipulable feature is the quantity of a particular good (and not the level of a liquid), etc. Such statements are not merely conventional truisms, nor do they depend on considerations of normal use, but rather derive from the causal properties of the mechanisms themselves, the kinds of loops they can form, and the subset that are loop-closing.

15.16.2 Semantic Value

Loop-closing theory reduces the concept of a 'true belief' to an information-bearing substate that (i) performs a belief-like functional role in a negative feedback control system (i.e., has world-to-mind 'direction of fit', combines with desire-like states to form actions, etc.) , and (ii) covaries with the control system's manipulable feature. A 'false belief', in contrast, performs a belief-like functional role and covaries with features of the world, but those features are not the manipulable feature.

Loop-closing theory avoids the disjunction problem because a control mechanism may be active in a loop —and hence instantiate substates that refer to the manipulable feature—without the manipulable feature being present. Representational error is possible because not all reliable causes of a belief-like substate are part of its extension. (I examine an extended example of misrepresentation in Sect. 15.17.3.)

15.16.3 Representation

The problem of semantic duality prompts the question: why does the winding of a thermostat's spiral coil represent temperature and not, more simply, itself? How does loop-closing theory exclude the 'opaque content' of information-bearing substates?

The problem of semantic duality arises because a given relation of reliable causation can participate in multiple kinds of control loops. For example, the thermal expansion of mercury can participate in the practice of measuring temperature (which selects the transparent content of the mercury column) or the practice of calibrating a thermometer (which selects the opaque content). Our goal-directed use of the variation of the height of the mercury column embeds it within an 'intentional circle' and fixes its content.

Loop-closing theory does not break out of the intentional circle but precisely specifies some simple kinds of causal structures that instantiate intentional circles. A thermostat, in sharp contrast to a thermometer, has causal properties, independent of our practices with respect to it, that instantiate a simple kind of intentional practice that manipulates temperature. In this kind of loop the spiral coil represents temperature (the transparent content) and not itself (the opaque content).

Simple cause-and-effect mechanisms, in contrast, such as thermometers, clocks, rulers, barometers, and accelerometers etc., do not possess semantic content independent of our practice with respect to them. They lack sufficiently rich causal structure to define a manipulable feature. In consequence, they merely possess 'derived intentionality'.

15.17 Applications of Loop-Closing Theory

Let's now consider some more elaborate applications of loop-closing semantics in order to further illustrate its explanatory properties.

15.17.1 Memory Addressing in CPUs

Sloman often points out, in response to claims that only human or biological minds possess original intentionality, e.g., Searle (1980), that 'there are clearly primitive semantic capabilities in even the simplest computers, for they can use bit patterns to refer to locations in their memories' (Sloman 2002b). It's true that CPUs use bit patterns in memory to point to other locations in memory. But it's not clear why Sloman considers this capability as a primitive case of original intentionality.

A typical von Neumann CPU architecture, in its essentials, comprises a Turing machine with finite memory and a read/write 'head'. Simplifying somewhat, a special register (some memory in the read/write head), called the Program Counter, points to the next instruction in memory. The CPU fetches the new instruction, by reading from memory, executes it, and then writes some new state to memory. Hence, a CPU performing a fetch/execute cycle is a kind of feedback control system with the local RAM as its environment. In philosophical discussions of Turing Machines the control aspect of computation is often overlooked (see Sloman (2002a) for a corrective).

The CPU executes the machine instruction

```
load reg1,(a)
```

that stores the value at memory address a in register reg1 (where (a) denotes the value stored at address a). Next the CPU fetches

```
add (reg1),5
```

and executes it. Let's stretch our terminology a little. Interpret 'add (reg1),5' as a desire-like state to add 5 to the contents of address 'reg1'. The belief-like state is the contents of address 'reg1'. The action-like state adds 5 to this value. The add instruction is successful if the content of address 'reg1' is updated. However, if 'reg1' points beyond addressable memory the CPU throws an exception and the instruction fails.

A full loop-closing analysis would require a detailed examination of the causal structure of the microprocessor's logic circuits. Nonetheless it's clear that 'local addressable memory' is part of the manipulable feature since the instructions fail if the CPU is not linked to RAM.

A CPU manipulates the contents of RAM. On a loop-closing account, therefore, the semantic content of the bit pattern stored at memory address 'a' is the addressable memory at '(a)', a conclusion that machine code programmers would find entirely underwhelming. The only additional insight loop-closing theory adds is that the semantic properties are reducible to the causal properties of the CPU's fetch/execute cycle. So this primitive semantic capability is not derivative on our practice with respect to the computer or our interpretation of its 'symbols'. For example, if the CPU happens to be executing a database program then the bits stored at address 'a' might refer, for users of the software, to an individual customer. But all this means is that the bit pattern at 'a' participates in multiple co-existing control loops or 'intentional practices'.

15.17.2 Dretske's Marine Bacteria

Dretske (1986) discusses the case of marine bacteria that 'have internal magnets (called magnetosomes) that function like compass needles'. The bacteria use their flagella to propel themselves forward while magnetic forces align their bodies parallel to the earth's magnetic field. In the northern hemisphere they tend toward magnetic north. Bacteria in the southern hemisphere, in contrast, have their magnetosomes reversed and tend toward magnetic south. In both hemispheres the magnetic field lines point away from oxygen-rich surface water to oxygen-free sediment at the bottom of the ocean.

Fodor (1997, Ch. 4) asks whether the magnetosome's alignment represents magnetic north or anaerobic conditions. Magnetic north 'carries information' or reliably covaries with the direction of anaerobic conditions. Hence, from the perspective of a 'crude causal theory', the semantic content of the magnetosome's alignment is indeterminate due to the conjunction problem.

Jacob adopts a teleological theory of the 'etiological function' of the magnetosome. An etiological function is an intrinsic causal power of a mechanism that explains its persistence through time. The magnetosome, as a magnetic material, intrinsically indicates magnetic north and this property confers survival advantage that was selected by evolution. Hence the magnetosome represents magnetic north.

Dretske (1986) notes a problem with this conclusion: misrepresentation now seems impossible. If we transplant a southern bacterium to the North Atlantic 'it will destroy itself—swimming upwards (toward magnetic south) into the toxic, oxygen-rich surface water'. Nonetheless the magnetosome, on the etiological account, is a veridical representation of magnetic north, despite the dire consequences for the organism.

Millikan's biosemantic theory identifies content with the conditions necessary for a mechanism to contribute to the 'proper function' of the whole organism (Millikan 1984). The proper function is fixed by evolution but, unlike the etiological approach, need not be an intrinsic causal power of the mechanism. We might claim that a magnetosome's proper function is to point the bacterium toward oxygen-free water. So a transplanted bacterium's magnetosome points in the 'wrong direction', misrepresenting the location of oxygen-free water.

My brief mention of these teleo-semantic theories of content cannot do them justice. I mention them only to draw contrast with the loop-closing approach. But, at first blush, it seems mistaken to locate semantic properties in the history of types of things rather than the causal properties of tokens of things. An exact functional copy of a marine bacterium, constructed in some futuristic lab, although lacking an evolutionary history should nonetheless have the same semantic properties as a marine bacterium in the wild (this is the standard 'swampman' objection to teleo-semantic theories (Davidson 1987); for further objections see Fodor (1990, Ch. 3)). As it happens, scientists in fact create artificial magnetotactic microorganisms (Kim et al. 2010).

Let's apply loop-closing theory to marine bacteria. First, a caveat: this is a philosophical thought experiment designed to test concepts and not a substitute for a detailed analysis of the biological design of a specific kinds of bacteria (e.g., certain species of marine bacteria navigate using a combination of magnetotaxis and aerotaxis (Bazylinski and Frankel 2004)).

The first question, from a loop-closing perspective, is whether a control mechanism (or mechanisms) recruits the magnetosome to perform a functional role as an information-bearing substate. Now the orientation of the bacterium is an entirely passive affair—it aligns north just like a compass needle (Chen et al. 2010). The magnetic force on the magnetosome 'biases' the bacterium's direction of motion but the magnetosome does not function either as a belief-like or desire-like state within a feedback control mechanism: the bacterium is simply blown about like a leaf in the wind. We merely have a 'one-way' cause-and-effect relationship between the mangetic field and the magnetosome. In consequence the orientation of the magnetosome, or the force it applies to the bacterium's body, does not map to any manipulable feature. The magnetosome is therefore like a mercury column in a thermometer rather than a spiral coil in a thermostat. Any semantic content it may have—such as pointing to magentic north or anaeorobic conditions—is derivative.

Loop-closing theory states that a magnetosome has no semantic content in virtue of the causal powers of the bacterium. This conclusion directly contradicts teleosemantic theories that, in comparison, look beyond the kind of thing the bacterium is and, in consequence, conflate the magnetosome's semantic properties with its possible adaptive functions on an evolutionary timescale.

15.17.3 A Fodor Machine

Fodor (1990, Ch. 3) introduces a helpful thought experiment to illustrate the problem of misrepresentation. You are in a park in the evening. You see a dog before you. It causes the symbol 'dog' to be instantiated in you mind, as part of a belief-like state B, 'there's a dog'. Later a furry, four-legged animal crosses your path, which also causes the instantiation of the symbol 'dog'. But—it turns out—the animal is in fact a cat. You were confused by the dim evening light. In this case, belief B is false.

Fodor explains that according to IS this is not a case of misrepresentation. Instead the symbol 'dog' refers to the disjunction *dog* or *cat-in-dim-light*, since the latter also reliably causes the activation of the symbol 'dog'. In fact, the problem is even worse than this, since *cardboard-dog-in-dim-light* and umpteen other 'dog-like' things might trigger your thought 'there's a dog'. How does loop-closing theory avoid the disjunction problem in Fodor's thought experiment?

Fodor's example is pitched at the level of conceptual content, whereas my account of loop-closing theory in this chapter is restricted to nonconceptual content. I wish to avoid the complication of the possibility of mismatches between a concept in an explicit ontology and the nonconceptual manipulable features that may be associated with it (a possibility that may explain how higher order false beliefs can cause control success). For simplicity, therefore, I will again avoid the complexities of the human mind and consider a rudimentary machine.

Imagine a 'Fodor machine', denoted $\mathcal{K}_1$, designed to scare dogs away. A camera generates input and a loud siren generates output. A software classifier maps the camera's input to a boolean B. We, as designers of the machine, intend $B = 1$ to indicate the presence of dogs, and $B = 0$ to indicate their absence. We train the classifier from data using some form of supervised learning and achieve a 5% false positive rate on the test set. The machine compares B against a constant value $D = 0$. If $B \neq D$ the machine sends the signal $A = $ 'on' to the siren, which starts blaring; if $B = D$ the machine sends the signal $A = $ 'off', which deactivates the siren.

The Fodor machine is a control mechanism with belief-like, desire-like, and action-like substates. As we hoped, when we test in the wild its loop-closing models include situations where dogs approach the camera, set off the siren and run away (where we experimentally control for dogs chasing nearby rabbits, cars honking horns, and so forth, such that the Fodor machine is the actual cause of its own success).

However, in dim-light, cats are a false positive for the classifier. In consequence, some loop-closing models include situations where cats approach the camera and get scared away. We designed $\mathcal{K}_1$ to be a dog-scaring machine but its manipulable feature turns out to be (at least) *dog* or *cat-in-dim-light*.

Loop-closing theory states that $\mathcal{K}_1$'s belief-like and desire-like states refer to this disjunction of features. In consequence if $B = 1$ when cats are in view this is not a case of misrepresentation, even though, from our perspective, the device is not working as intended.

Now we place a cardboard cutout of a dog before the camera. In dim-light the cardboard dog is also a false positive for the classifier. The siren starts blaring but, of course, the 'dog' does not move. This world is not loop-closing. Machine $\mathcal{K}_1$ is active—it 'tries' to scare the cardboard dog away—but the manipulable feature is absent, and therefore control success, i.e., a state of inactivity, is not attained. $\mathcal{K}_1$'s manipulable feature does not include *cardboard-dog-in-dim-light*. In consequence if $B = 1$ when the cardboard dog is in view this *is* a case of misrepresentation because B falsely represents the presence of the manipulable feature, viz. *dog* or *cat-in-dim-light*.

The point is this: some mechanisms may be highly dedicated systems with univocal focus; others may successfully control a wider collection of disparate things and therefore possess a manipulable feature expressed as a disjunction of features in a metalanguage. Some semantic indeterminacy should be expected. For example, a necessary condition for cuckoo brood parasitism is that both real chicks and cuckoo chicks are loop-closing models for the host bird's control mechanisms. However, this kind of indeterminacy does not constitute a disjunction problem in loop-closing semantics because error is always possible: a reliable cause of a belief-like substate (e.g., *cardboard-dog-in-dim-light*) need not be part of its extension (e.g., *dog* or *cat-in-dim-light*). As Fodor (1990, p. 59) remarks 'the least you want of a false token is that it be caused by something that is not in the symbols extension'.

15.17.4 Ramsay-Whyte Success Semantics

The loop-closing account of the semantic value of beliefs shares common features with the philosophical theory of Ramsay-Whyte success semantics (Whyte 1990, 1991), which is 'an heir to the pragmatist tradition' (Blackburn 2010). Success semantics is pitched at the level of abstraction of the propositional attitudes and—simplifying—states that 'the truth of beliefs explains the success of the actions they cause' (Whyte 1990). The key idea is that actions succeed if our beliefs represent the world correctly. For example, I believe the animal before me is a dog. This belief may be true or false. I decide to make the animal bark by shouting and running around it. If the animal really is a dog then, all other things being equal, I succeed; otherwise (say, it's a cat on a dim evening)—I fail.

Crane (1995) summarizes success semantics (SS) as

> (SS) A belief-like state B is a veridical representation of state β if and only if actions A caused by B and desire-like state D would succeed if β obtained.

The relation to loop-closing semantics is clear. Both approaches reduce semantic properties to the 'practical' or pragmatic consequences of belief-like, desire-like and action-like states (i.e. episodes in which an agent attempts to 'control' its environment).

A common criticism of SS is that 'success' is a semantic property that it fails to reductively define (e.g., Crane (1995, p. 189)). SS defines 'success' as the satisfaction of a desire, which is the bringing about of the desired state-of-affairs. So what a belief represents is defined in terms of what a desire represents. But this just postpones the explanation of representation. SS is therefore a circular, not a reductive, explanation of semantic properties (see Whyte (1991) for counterarguments that attempt to avoid this conclusion).

Loop-closing theory is different in this respect. It ultimately defines 'success' in terms of a state of inactivity (see Definitions 15.5, 15.6 and 15.11). Action-like states, unlike beliefs and desires, are *not* representational states, or at least certainly not when the 'rubber hits the road' and the action is performed. Actions may have imperative semantics and ultimately be effective or ineffective with respect to the beliefs and desires that cause them. But such semantic properties are irrelevant to the determination of whether an action is performed or not. von Wright (1971) puts the matter nicely: 'The connection between an action and its result is intrinsic, logical and not causal (extrinsic). If the result does not materialize, the action simply has not been performed. The result is an essential "part" of the action'. The performance or nonperformance of an action is therefore a kind of event that can be defined and identified without mention of semantic properties. 'Success', in loop-closing semantics, is defined reductively in terms of inaction. We'd hope that a causal account of semantic properties would ultimately 'ground' reference in terms of actions—and this is precisely what loop-closing semantics achieves.

15.18 Conclusion

Sloman's sketch of loop-closing semantics aims to reductively explain a subset of semantic properties in terms of the causal powers of generalized control systems interacting with their environments. In this chapter, I have developed Sloman's sketch, in particular by specifying a metalanguage to describe the causal structure of loop-closing models and identifying a control mechanism's manipulable feature. The manipulable feature is that subset of the world necessarily present for control success, where 'success' is a control episode in which the mechanism is the actual cause of its own inactivity. The manipulable feature is the key concept that enables loop-closing semantics to explain semantic reference, focus, and value.

Thermostats are mundane artifacts that might not seem worthy of philosophical reflection. But simple control systems encourage us to critically re-examine the (often pre-theoretic) concepts we employ to specify and describe semantic properties, such as 'refer', 'represent', 'believe', 'desire', etc. Sloman (1994a) writes: 'I claim that there are no well-defined concepts that correspond to our normal use of these words. Rather there are many different features to be found in the contexts in which we ordinarily talk about meaning, and different subsets of those features can occur in connection with control states of various kinds of machines'.

Control systems, such as thermostats, are qualitatively different from passive systems, such as thermometers. Thermostats possess semantic properties in virtue of their causal powers; in contrast, thermometers possess semantic properties in virtue of our intentional practices. The causal structure of a simple control system is sufficient, and perhaps necessary, to instantiate semantic properties. In consequence, simple control systems, such as the humble thermostat, constitute a fundamental building block from which to construct a general, naturalized theory of semantics.

Acknowledgments My thanks to Aaron Sloman whose ideas and general approach are of course the main inspiration for this chapter. Thanks also to Andrew Trigg, Brian Logan, and attendees of the 2011 symposium in honour of Aaron Sloman, held at University of Birmingham, who provided useful feedback on an earlier version of this chapter.

References

Barwise JK, Seligman J (1997) Information flow: the logic of distributed systems. Cambridge University Press, Cambridge

Bazylinski DA, Frankel RB (2004) Magnetosome formation in prokaryotes. Nat Rev Microbiol 2:217–230

Bhaskar R (1997) A realist theory of science. Verso Classics, London. (Original edition published by Leeds Books Ltd 1975)

Blackburn S (2010) Success semantics. Oxford University Press, Oxford, pp 181–199

Chen L, Bazylinksi DA, Lower BH (2010) Bacteria that synthesize nano-sized compasses to navigate using earth's geomagnetic field. Nat Educ Knowl 1(10):14

Crane T (1995) The mechanical mind, 2nd edn. Routledge, London

Davidson D (1987) Knowing one's own mind. Proc Addresses Am Philos Assoc 60:441–458

Dennett DC (1997) True believers: the intentional strategy and why it works. In: Haugeland J (ed) Mind design II: philosophy, psychology, artificial intelligence. Massachusetts Institute of Technology, Massachusetts. (Reprint of a 1981 publication)

Dretske F (1981) Knowledge and the flow of information. MIT Press, Cambridge

Dretske F (1986) Misrepresentation. In: Bogdan R (ed) Belief. Clarendon Press, Oxford

Feferman AB, Feferman S (2004) Alfred Tarski, life and logic. Cambridge University Press, Cambridge

Fodor J (1989) Psychosemantics: the problem of meaning in the philosophy of mind. MIT Press, Cambridge

Fodor J (1990) A theory of content and other essays. MIT Press, Cambridge

Jacob P (1997) What minds can do. Intentionality in a non-intentional world. Cambridge studies in philosophy. Cambridge University Press, Cambridge

Kim DH, Cheang UK, Köhidai L, Byun D, Kim MJ (2010) Artificial magnetotactic motion control of tetrahymena pyriformis using ferromagnetic nanoparticles: a tool for fabrication of micro-biorobots. Appl Phys Lett 97:17302–17303

McCarthy J (1979) Ascribing mental qualities to machines. In: Philosophical perspectives in artificial intelligence. Humanities Press, Atlantic Highlands, pp 161–195

McCulskey EL (1959) Minimization of boolean functions. Bell Syst Tech J 35:149–175

Millikan RG (1984) Language, thought and other biological categories. MIT Press, Cambridge

Pearl J (2000) Causality, 2nd edn. Cambridge University Press, Cambridge

Searle J (1980) Minds, brains and programs. Behav Brain Sci 3:417–457

Sloman A (1986a) Reference without causal links. In: European conference on artificial intelligence, pp 191–203

Sloman A (1986b) What sorts of machines can understand the symbols they use? Proc Aristotelian Soc Suppl 60:61–95

Sloman A (1994a) Representations as control sub-states. Cognitive Science Research Centre. School of Computer Science, University of Birmingham, UK

Sloman A (1994b) Semantics in an intelligent control system. Philos Trans R Soc Phys Sci Eng 349:43–58

Sloman A (1997) Beyond Turing equivalance. In: Millican P, Clark A (eds) Machines and thought: the legacy of Alan Turing, vol 1. Oxford University Press, Oxford

Sloman A (2002a) The irrelevance of turing machines to artificial intelligence. In: Scheutz M (ed) Computationalism: new directions, MIT Press, Cambridge, pp 87–128

Sloman A (2002b) The mind as a control system. In: Hookway C, Peterson D (eds) Proceedings of the 1992 royal institute of philosophy conference 'philosophy and the cognitive sciences'. Royal Institute of Philosophy, Cambridge University Press, Cambridge

Tarski A (1956) Logic, semantics, metamathematics: papers from 1923 to 1938 by Alfred Tarski. Clarendon Press, Oxford

von Wright G (1971) Explanation and understanding. Cornell University Press, Ithaca

Whyte JT (1990) Success semantics. Analysis 50(3):149–157

Whyte JT (1991) The normal rewards of success. Analysis 51(2):65–73

Wiener N (1975) Cybernetics: or control and communication in the animal and the machine, 2nd edn. MIT Press, Cambridge

Woodward J (2003) Making things happen. Oxford University Press, Oxford

Index

J. L. Wyatt et al. (eds.), *From Animals to Robots and Back: Reflections on Hard Problems in the Study of Cognition*, Cognitive Systems Monographs 22,
DOI: 10.1007/978-3-319-06614-1, © Springer International Publishing Switzerland 2014